FARM OFFIC
HANDBOOK

2nd Edition

FARM OFFICE HANDBOOK

2nd Edition

IAgSA Institute of Agricultural Secretaries & Administrators

First published 2012 by Old Pond Publishing Ltd

Second Edition
Published 2017 by

Old Pond Publishing,
An imprint of 5M Publishing Ltd,
Benchmark House,
8 Smithy Wood Drive,
Sheffield, S35 1QN, UK
Tel: +44 (0) 0114 246 4799
www.oldpond.com

In association with

IAgSA
East Haddon Lodge
East Haddon
Northamptonshire
NN6 8BU
Email: IAgSA@IAgSA.co.uk

A catalogue record for this book is available from the British Library

ISBN 978-1-910456-57-6

Book layout by Servis Filmsetting Ltd, Stockport, Cheshire
Printed by Replika Press Pvt Ltd, India
Photos by IAgSA unless otherwise indicated

DISCLAIMER

All reasonable steps have been taken to ensure that the information in the *Farm Office Handbook* is correct. However, the correctness and completeness of material within it is not guaranteed. Website addresses are included throughout the book to help readers to check details for themselves. Neither the Institute nor any contributor is liable or responsible for any kind of loss or damage that may result to the reader or a third party as a result of the use of this book.

Contents

Acknowledgements

The *Farm Office* was first published in 1976. It quickly became the bible for farm administrators in the UK, being a valuable source of knowledge for those working in remote farm offices. With the internet today, our knowledge sources are only a few clicks away, provided you know where to look. This revised handbook has therefore been written as a first port of call for rural administrators, covering a wide range of tasks that may need to be completed. In addition, it contains useful links to websites where up-to-date information regarding current legislation can be downloaded. When Article 50 of the Lisbon Treaty is invoked and the process of leaving the EU commences, readers are advised to use internet searches to ascertain the new procedures.

The purpose of the Institute of Agricultural Secretaries and Administrators (IAgSA) is to support, train and promote the profession of farm and rural administrators. The handbook is a useful resource in achieving all three of those aims but in particular as a learning tool to support the IAgSA training programme.

The role of a good farm or rural business administrator requires the ability to cope under pressure and to keep up-to-date with ever-changing rules and regulations. It is also essential to know where to find current information but also to have the background understanding and underpinning knowledge related to farm office activity. As chairman of IAgSA, I am therefore very proud that we have had the opportunity to produce the second edition of the *Farm Office Handbook*, which has been, and will continue to be, such a valuable asset to the industry.

I would like to acknowledge the work done by the team who have updated this handbook, and especially Jenny Pine, who has coordinated the work.

Finally, we are very grateful to the Chadacre Agricultural Trust for its continued support, which has enabled us to produce this revised edition of the *Farm Office Handbook*. The Chadacre Agricultural Trust is a charity: your legacy will help ensure that the Trust continues to achieve its aims of supporting education and training for farming in East Anglia. www.chadacre-trust.org.uk

Cathy Meredith MBE ARAgS
FIAgSA IAgSA Chairman

Co-authors

Jenny Pine
Cathy Meredith
Ann Davies
Tracey Nicholls
Sally Lemonius
Tim Cartwright

With support from

Meg Cowap
Elin Jones
Billie Johnson
Nancy Sloan

Foreword

As patron of the Institute of Agricultural Secretaries and Administrators (IAgSA), I am pleased to be writing the foreword of the second edition of *Farm Office Handbook*. J.D. Alston in the foreword of the original *Farm Office* by Michael Hosken, wrote how this 'fills a gap in farming literature' and this remains true 40 years later. This handbook is a useful addition to any farm office, as it gives a complete overview of what is expected from the administration of a farm, from managing accounts to dealing with office logistics. It provides a consistent approach to how a farm office should be run, which is crucial irrespective of the size of the farm business.

The handbook should serve as a checklist to ensure farm businesses keep good records – from financial accounts to those that show their compliance with statutory regulations. It is also an important reference tool for those starting out as a farm/rural administrator, and those who have been in the business for some time.

The role of the farm administrator is invaluable in ensuring the farm office and the farm run smoothly. Farm administrators need to have the information at their fingertips (invoices, delivery notes and so on) and the ability to process, extract and provide information to the management team to enable them to make informed husbandry and business management decisions.

This handbook guides the reader through a range of areas; from farm office basics to planning annual budgets for the forthcoming year – a practical approach to running an effective and efficient rural business.

Sir Jim Paice
Minister of State for the Department for Environment,
Food and Rural Affairs, 2010–2012

Chapter 1

The Farm Office

Agricultural and Allied Businesses

In recent years, there has been a shift towards specialisation and larger farming operations. A well-resourced farming business may provide labour, machinery and expertise to neighbouring smaller farms, benefiting all parties from economies of scale. Farm diversifications are now commonplace – for example, let property, environmental schemes and non-food crops – providing important additional income and, generally, more paperwork. Whatever the size and nature of the business, there will be a farm office.

Farm Office Handbook is intended as a practical reference source. We look at setting up straightforward accounting and recording systems for farming and allied businesses, using spreadsheets and computerised systems. Areas such as VAT and payroll processing are covered in detail. However, legislation is far-reaching and the structure and scope of a business will determine what legislation is applicable, therefore we aim to highlight legislation that may apply and point the reader to the appropriate website for up-to-date information. Website addresses used in the book are current at the time of going to press but if a particular website cannot be located – for example, due to government departmental changes – try using a search engine for the topic concerned.

Farm administrators come in two varieties:

1. **Professional:** The first category might be the professional who has come to specialise in farm administration following work experience in accountancy or other disciplines in industries either related

or unrelated to land-based enterprises. The greatest proportion of UK land is farmed in large holdings, owned or rented, perhaps from estates of even greater size. Many such businesses require full-time administrators, qualified and skilled in the particular range of expertise that characterise agriculture and horticulture, possibly with equestrian, tourist or other appendages. An alternative career choice may lie in self-employment, servicing a number of clients. See Chapter 15.

2. **Family members managing farm records and accounts:** Whereas most UK farmland is in large units, the remainder is made up of a huge majority of smaller holdings. In terms of enterprise size, a very small percentage of holdings employ more than five people. So who 'does' the farm accounts and recordkeeping? This traditionally falls to the proprietor or a family member.

Farm accounts and recordkeeping falls into two main categories:

1. **Statutory:** Compliance with the requirements of statutory regulations and standards of farm assurance schemes. These include:

   - ***Financial records***, particularly those regulated by HM Revenue & Customs (HMRC) including Value Added Tax (VAT), tax assessment on profits, wages (also involving other legislation) and Pay As You Earn (PAYE).
   - ***Physical records*** regulated by other government departments such as Department of the Environment, Farming and Rural Affairs (DEFRA), Natural England; Environment Agency (EA), Health & Safety Executive (HSE), Food Standards Agency (FSA) and the Office of Fair Trading (OFT).
   - ***Evidence of compliance*** with statutory regulations involved in farm assurance schemes – voluntary schemes that producers can join to assure customers that certain standards have been maintained in the production process, such as 'organic' and 'free range'.

2. **Management:** Management records and accounts will depend upon the nature of the business but may include:

   - ***Financial records*** – to facilitate control of cash flow, and comparative analysis year-on-year and with national benchmarks.

- ***Physical records*** – monitoring crop and animal husbandry, pedigree records, etc.
- ***Stock records*** – crops in store, seed, fertiliser, crop protection products, feed stuffs, fuel, machinery parts, etc.
- ***Property records*** – records of condition, inventories, etc.
- ***Machinery records*** – service records, etc.

The Basis of Good Recording

Basic to any business is recording details of day-to-day (or month-to-month) financial transactions. The form in which they are recorded must comply with several principles:

- Amounts must be accurate and the layout must lend itself to cross-checking.
- Analysis must be used to reveal information useful to management.
- The system must ensure that there are no omissions.
- Full values of goods and services must be shown, even in cases of part exchange where the cash sum represents only the price difference.
- It must be possible to keep data in step with what is happening to the bank account.
- It must be easy to extract information required for VAT returns.
- The eventual totals should, as far as possible, be in a form suitable for transfer to the annual accounts.
- Where a family farm is run on a single bank account, it is necessary to be able to distinguish between business and private transactions.
- Provision may need to be made for the main accounts to be integrated with the wages book, petty cash account and any separate capital, loan or reserve account.

Choice of Accounting System

A great deal of thought and expertise has gone into devising 'user-friendly' systems suitable for farming and other small rural businesses. The choices are:

- **Cashbook, single-entry systems:** Computer spreadsheets based on the traditional cash analysis book are easy to set up and use. Paper-based systems are still widely used – a variety of forms are available commercially, pre-printed or blank, in book form or loose-leaved. See Chapter 4.

- **Computerised accounting programs:** These can be either general or agriculture-based. There is a choice of single-entry system, like the analysed cash book, or double-entry simulating a ledger system, where moneys owed to and by the business may be tracked and an asset register maintained. Refer to Chapter 6 for more information.

Taking the Accounting System a Step Further

Most businesses now have to file statutory reports online such as VAT returns and payroll information. This may be an incentive to computerise accounting and payroll records. See Chapter 5.

Whichever system is in place, the primary aim is to produce a profit and loss account and balance sheet after 12 months of trading. Where a computerised accounting programme is used to its full extent, it should be possible to produce final accounts having entered the valuation and made manual adjustments if necessary to the opening balance. See Chapter 6.

The Profit and Loss Account

This is a summary of the working of the business during the past year. It takes account of:

- money paid out and received
- debts owed to and by the business
- values of the business assets (livestock, growing crops, stocks) at the beginning and end of the year and changes (depreciation) in the case of machinery.

Refer to Chapter 9 for more information.

The Balance Sheet

Whereas the profit and loss account covers a period, the balance sheet is more like a flash-photo: it shows the state of the business at any one time, normally the end of the financial year.

It lists everything the business has, its *assets*:

- crops and animals
- machinery
- land
- credit balance at bank
- debts owed to the business
- cash in hand.

Against which are set its *liabilities*:

- mortgage
- loan
- overdraft
- debts owed by the business.

To reveal the *net worth* of the business at the specified date. See Chapter 10.

Purchase and Sales Documentation

Although the accounting side of farm office work is the aspect that springs most obviously to mind, many items of paperwork may precede the final transaction. Each document performs a specific function, and the farm administrator must know not only what to do with each one, but also which ones may be safely thrown away and when. See Chapter 3.

Budgeting and Management Decisions

The essence of planning is to ask, 'What would happen if…?' Changes must be investigated before being put into effect. Finances must be worked out in the form of a budget before investments are made. The farm administrator must be able to extract data and present information relating to plans of management for a year, or maybe for a longer period, depending upon the nature of the planned changes. Enterprise budgets must then be supported with cash flow budgets.

An annual budget is produced for the forthcoming year, and the farm administrator plays an important role in providing some of the information on which the necessary informed decisions can be based. If no major changes are to be made, such as a new enterprise or major capital expenditure, then overheads will remain in line with inflation and reference may be made to the preceding year's accounts as a starting reference point. However, the cropping plan, livestock numbers, market trends and timing of sales will have an effect on income and variable costs and this information is more likely to come from the farm manager. Refer to Chapter 11 for more information.

The decisions at which management arrives may not be the obvious ones since other matters besides information may be relevant. There may be limitations or pressures not revealed by the paperwork; for example, increasing age may limit ambition, or an interest and preference for one type of husbandry may outweigh the budgeted advantage of a different system.

Gross Margin Analysis

The concept of gross margins is established in farm management accounting; it simplifies the multitude of farm sizes and types into a 'unit of production basis' – per hectare or per head. Gross margins are used to compare performance, year on year within a business or with nationally published figures based on farm surveys. The profitability of different enterprises within a farm may be compared using gross margins; therefore they are useful in budgeting. Computerised farm recording systems generally have a dual purpose; for example, cultivations, seed, fertiliser and spray applications must be recorded for cross-compliance purposes. With the addition of harvesting records and sales, crop input costs and other variable costs, a gross margin can be produced. It is important for the administrator to be conversant with this concept as it is widely used: further information is found in Chapter 11.

Other Measures of Physical Performance

Computerised information systems are now widely used in large specialised units and generally the unit manager will be responsible for the data input, interpretation and reporting rather than the farm administrator. Therefore, commonly used ratios are explained in Appendix 2 with pointers to specialist websites where more detailed and up-to-date information can be found.

Other Farm Records

While the pattern of accounting procedure is basically similar on any farm, the choice of husbandry records depends not only on what enterprises exist on the farm, but also on their intensiveness and membership of quality assurance schemes. So the extent of the farm administrator's involvement in enterprise records is very variable indeed. See Chapter 12.

Working with Other Professionals

As the agricultural business environment becomes more complex and the implications of falling foul of legislation expensive, the need to engage the services of specialists increases. Most businesses use the services of accountants to finalise accounts at the end of the financial year and deal with tax returns. It makes sense for the administrator to establish a good working relationship with the accountant. Generally they will encourage and help the administrator to take the accounts to a sufficiently high level, so that they can spend their time on dealing with taxation rather than bookkeeping.

Land agents and farm management consultants are widely used for various matters, such as, for example, the annual farm valuation and completion of government forms in connection with farm grants and subsidies. However, these particular tasks are often dealt with 'in-house'. Contractual arrangements such as tenancies and contract farming agreements are usually drawn up by an agent where the obligations of both parties are set out in detail. Gentlemen's agreements are still used; they rely upon the honour of the parties for their fulfilment, rather than being in any way enforceable.

With the increase in complexity and constant revision of health and safety legislation and employment law, HR (human resources) consultants are sometimes used, particularly where there are a number of employees. Similarly, health and safety (H&S) specialists may be retained on a contractual basis to carry out initial risk assessments. They will implement any immediate action required, put in place the correct H&S documentation and work with the business to keep it up-to-date.

The main membership organisations supporting rural businesses are the National Farmers Union (NFU), Country Land and Business Association and the Tenant Farmers Association. The equivalent bodies in Scotland are: NFUS and Scottish Land & Estates and the Scottish Tenant Farmers Association. Helplines and website details can be found in Appendix 1.

Farm Office Equipment and Office Organisation

The Basics

A farm office, just like any other, should be a comfortable and safe place in which to work. The basic requirements are a desk and swivel chair, filing cabinets and a lockable cupboard with shelving. Electrical power and telephone points are also essential.

A desk must have sufficient surface space for a computer monitor, keyboard, telephone and a stacking tray system where incoming documents can be sorted and allocated. Lockable desk drawers are advisable where cheque books etc. can be kept safely, together with a filing drawer containing suspension files where documents in regular use, such as bank statements and payroll, can be stored. It may also be advisable to install a small fireproof safe where cash, card readers, data backups (memory sticks) and other important items can be safeguarded.

Filing Systems and Archiving

A filing system for semi-permanent storage is necessary plus storage for archives. Accounts and VAT records should be boxed, labelled clearly with the accounting year dates and kept for six years. Payroll records should be kept for the current and previous three years.

Some physical records may need to be kept to support RPA payments: for details of the relevant legislation, consult the guidance notes for the particular scheme.

Unless it is a new business, there is likely to be a filing system in place that may have evolved over some years. It is likely to be a combination of the following commonly used systems and, if it works, it is probably sensible to work with it and make improvements where possible.

Commonly Used Filing Systems

- **Alphabetically (lever arch files):** Useful for purchase invoices in a small business with no need to refer to an accounts program for a number.
- **Numerically (lever arch files):** Sales and purchase invoices, number generated by accounts program, may be subdivided into months.
- **Chronological (suspension files):** Suitable for grant/subsidy claims, crop records by year.
- **Grouped numerical:** Allocate a letter to a subject area, and then subdivide by number, for example:

  A = Finance: A1 Mortgage, A2 Bank Loan.
  B = Professionals: B1 Accountant, B2 Bankers, B3 Land Agent.
  C = Legislation: C1 Health & Safety general, C2 Waste Exemptions.
  D = Insurance: D1 Vehicles, D2 Farmers Combined Policy.

Data Security and Backup

Regular 'backups' should be made of all data as recommended by the software provider. There are various options such as a separate disc drive specifically for backups. Automatic backup facilities are also available in some internet security programs. Additional backups should be made on a memory stick and kept off-site.

Many agricultural software providers also offer hardware and software support contracts and a remote data backup facility via the internet. They will also recommend a suitable internet security program. Alternatively, it is possible to source these services and advice from independent computer consultants. Having a contract in place with a specialist who is easily accessible gives valuable peace of mind to the farm administrator, who may be regarded as the in-house IT department as well as everything else!

Basic Equipment – Tips

- *Calculator* – choose a simple calculator with a large display.
- *Computer* – PCs and laptops with wireless accessories and large screens are recommended.

- *Printer/scanner/photocopier* – check cost of ink cartridges when buying a printer, laser printers may be more economical for high usage.
- *Shredder* – useful for instant disposal of unwanted confidential information.

Computer Software Packages

General Administration – Microsoft Office

- Word documents – general correspondence, etc.
- Excel spreadsheets – calculations, presentation and sorting of data.
- Outlook – emails, maintaining and sorting contact lists, diaries and 'to-do list'.
- (NB. Task manager is very useful for deadline reminders)

Financial Accounts

Accounting Software Package – see Appendix 1:

- Specialist agricultural
- Commercial
- Cloud-based systems.

Payroll

- Commercial software
- Cloud-based systems
- Free payroll 'Basics Tools' available from HMRC, www.gov.uk.

Management Accounts and Compliance

- Crop and Field Recording Software Package (and cross compliance). See Appendix 1.
- Livestock Recording Software Package (and cross compliance). See Appendix 1.

'Do It Online'

An individual, business or agent must be registered with Government Gateway (www.gateway.gov.uk) before making use of the many online services available through HMRC and DEFRA. It is only necessary to register once; the various services can be selected on the Government Gateway website after completing the registration process. A password is required comprising letters and numbers. Following the registration process a user ID will be issued via email and online services will then be available for use.

Using the Internet

Use the internet for up-to-date information. The websites of all government departments and many other agencies and public bodies can all be found at www.gov.uk. EU and other legislation constantly changes, therefore it is most important to check facts on the www.gov.uk website.

Top Tips

- Establish a regular routine and keep on top of paperwork.
- Keep up-to-date with rules and regulations with email alerts (from HMRC, EA, etc.) and bulletins (from IAgSA, NFU, etc.).
- Filing online is easy, once a computerised system is in place.
- The accounting system's main function is to record financial information.
- Use a separate system for recording detailed physical information.

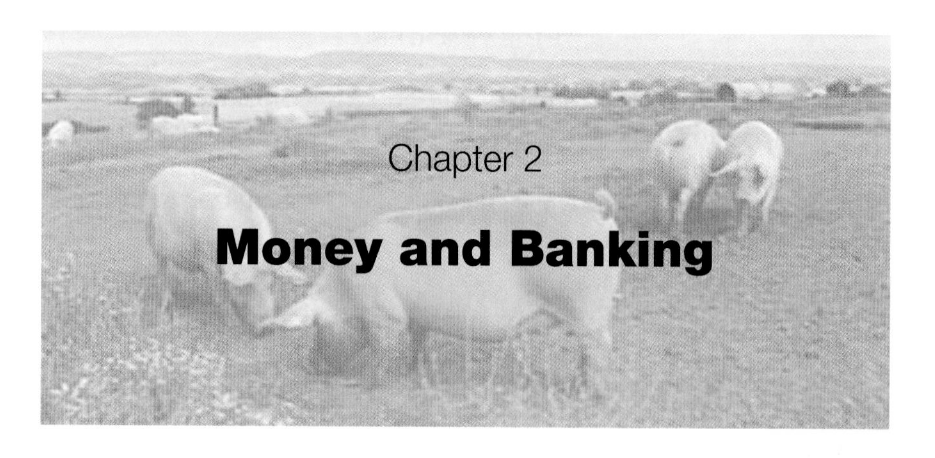

Chapter 2

Money and Banking

Like any business, farming uses capital and labour to generate profits. Resources such as land, machinery, seeds and fertilisers are worked to produce the food. However, the most convenient way of thinking of the value of the land is in money terms. Similarly, money is needed to acquire labour, machinery and inputs.

In farming, the land, farmhouse, farm cottages and permanent farm buildings with their fixed equipment such as grain bins and milking equipment are regarded as being permanent resources. A landowner, whether a private individual or a corporate body such as an insurance company, can either use that land by farming as owner-occupier (using salaried managers if necessary) or let it for rent to some other private individual or farming company to farm as tenant.

The tenant, however, will need considerable money of his own in order not only to pay the rent, but also in greater amounts to buy livestock and dead-stock (machinery, temporary buildings and so on), seeds and fertiliser, to pay wages to any employed workers and to keep himself and his family until such time as the farm generates sufficient income.

Cash transactions for the purchase of goods have been carried out for thousands of years. However, now it is used less and less. Most of the day-to-day and long-term financial transactions will be made via a bank account. The increasing use of online banking saves time, and another useful function is the ability to view statements as any time. In this chapter we give an overview of traditional banking facilities, other methods of paying for large capital items of expenditure such as machinery, and simple methods of accounting for petty cash.

Banking Transactions

Banks now offer many different ways of making and receiving payments, for both personal and business purposes.

Cheques were the most widely used form of banking transaction and still are the traditional paper-based method of making a payment from the bank account of the *drawer* to the bank account of the *drawee.*

Electronic Transactions

Electronic transactions are becoming the preferred way of conducting banking business.

Online Banking

Online banking (or internet banking) allows a business to conduct financial transactions on a secure website operated by their bank.

Online banking facilities have many features and capabilities in common, but traditionally also have some that are application-specific. The common features fall broadly into the following categories:

- Account-to-account transactions ('wire transfers').
- Transfers between 'our' accounts: current, loan, deposit, savings, personal, etc.
- Transactions to and from other business accounts in settlement of invoices, etc. ('direct credits', 'direct debits'). This may include the purchase and sale of investments, phone top-ups, and so on.

Account management:

- Statements of account for revealing amounts of interest, bank reconciliation, etc.
- Setting up new standing orders or direct debits.
- Opening new accounts.
- Application for overdraft facility changes ('credit limits') or loan.

Banks vary in the available online facilities, some allowing authorised non-signatory personnel to view statements and other bank information without being able to process a transaction.

Authorisation and Security

Whereas traditional banking security was based on personal identification and signatures, online banking relies on PINs (personal identification numbers), passwords and security devices. Where a bank account has one authorised signatory, the account-holder, and no online facility, it is a straightforward situation where the account-holder signs all cheques and authorities. However, banking online is now commonplace in the farm office and a procedure for the authorisation of transactions needs to be put in place. Passwords and card-reading devices must be kept securely as per the instructions of the bank.

Self-employed administrators and bookkeepers with authority to carry out transactions are advised to have something in writing from the client, recording your responsibility and the actions you are permitted to undertake.

Prevention of Cybercrime Attack

Small businesses are prime targets for cybercriminals. Generally their security is less robust than a large company that invests time and money on IT systems. In many cases it is months before a small business discovers that the bank account has been hacked. Social engineering is the manipulation of situations and people to gain access to secure systems and information and requires minimal technical knowledge. Things to look for:

- **Email – phishing:** Even though the sender's name might sound legitimate, ask: is the subject line alarmist with excessive punctuation? Is the logo low quality? Is time of sending unusual? If so, do not click on any link or open any attachments.
- **Phone calls – vishing:** Never give out confidential information (PINs, passwords, ID numbers etc.).
- **Texts – SMiShing:** Never confirm personal or financial data or supply account information.

Remember, banks or reputable organisations will never ask you for your password or PIN via email or phone call. If you think someone knows it, change it immediately.

- **Cheques:** Never pay refunds to somebody against uncleared funds.

Useful Tips for Cyber Security

- Have a strong password – use a mixture of lower- and upper-case letters, numbers and non-alphanumeric characters.
- Have malware protection.
- Make sure you are on a secure webpage by checking it is 'https'.
- For home or mobile working – protect data using an appropriately configured virtual private network. Visit www.getsafeonline.org.

Other Banking Options

- **Standing orders** are instructions to a bank to pay a set amount at regular intervals from one bank account to another.
- **Direct debits** are instructions that a bank account holder signs in favour of a beneficiary ordering his/her bank to accept a request from that beneficiary's bank to pay an amount directly from his/her account. It is more flexible than a standing order because it is often worded in such a way that it accommodates a change in the amount or payment frequency. It is similar to a direct deposit or direct credit but initiated by the beneficiary.

Payment Cards

Debit Cards

Debit cards (also known as bank cards or cheque cards) provide an alternative payment method to cash when making purchases. Functionally, they can be called electronic cheques, as the funds are withdrawn directly from either the bank account or from the remaining balance on the card. Debit cards may also be used for instant withdrawal of cash from cash point ATMs (automated teller machines) and as cheque guarantee cards. Shops may also offer cash-back facilities to customers, where cash may be drawn along with purchases. Many shops, cafes etc. now have the ability for the cards to be read electronically without contact with the machine (*contactless payments*) and without the use of a PIN number. There is usually a low transaction limit on these.

Credit Cards

Credit cards allow the card-holder to buy goods and services based on the holder's promise to pay for these goods and services. The issuer of the card creates a revolving account and specifies a maximum amount of credit up to which the user can borrow money for payment to a merchant or as a cash advance to the user. Credit cards allow consumers a continuing balance of debt, subject to interest being charged and a minimum monthly repayment being made, though no interest is charged if the previous month's balance is cleared within a specified time limit.

Pre-Paid Cards

Pre-paid cards are 'topped-up' by transferring money into the pre-paid card account. They can then be used as a normal cash or credit card; however, once the money is spent, there is no facility for creating a debt. They are available to small businesses and may be a useful means of controlling business expenditure by employees.

Telephone Banking

This is a facility that allows customers to make account enquiries and perform transactions over the telephone.

Deposit Accounts

Accumulated cash reserves will earn minimal or no interest if left in a current account. A business reserve account linked to a current account with an automatic transfer facility between the accounts may be useful and will earn a slightly higher rate of interest. Some deposit accounts may be subject to a period of notice to the bank, and withdrawals can be made only by credit transfer into the current account. For a better rate of return on a large sum of money that is not needed in the short term, treasury deposit or money market deposit accounts may be recommended by the bank manager. In general, a higher rate of interest will be paid on a longer fixed term. Such interest has to be accounted for as part of the farm income and is taxable, just as bank interest charged on mortgage or business loan is a business expense and therefore usually allowable against tax.

Euro Accounts

Some farmers elect to have farm support grants paid in euros or they may have dealings with countries in the 'Eurozone'. A euro account is available from most banks but usually only for their existing customers. The account allows the farmer to hold payments received in euros in that currency, rather than having the payments converted into Sterling and placed into a Sterling account. This can be useful if the farmer has to make outward payments in euros at a later time. However, if the farmer has no such requirements, there may be little benefit in holding the funds in euros at all.

Borrowing Money

As very few businesses are able to generate funds without a prior need to spend some money, any individual wishing to start in business should ideally have some money of their own. Any additional requirement will have to be borrowed. It could, in theory, be obtained by 'going public' – raising capital through stocks and shares – although this is very rare in farming as it applies only to the largest of companies. It is much more common for a business to have to borrow money from a bank.

Banks very often seek 'security', which in this case means collateral, i.e., goods that the bank can acquire a legal right to sell, if necessary, to repay the borrowing if the business cannot meet the costs. The more security that can be offered, the easier it is to borrow money. Having an appreciable financial stake in the venture oneself is helpful, as this will provide a 'safety margin' that can be used during difficult trading circumstances and will lessen the degree of risk that the bank is asked to take.

Another vital consideration for any bank being asked to support a business is its ability to repay the borrowing in accordance with the agreed programme. A bank will require some projections from the business, usually both for cash-flow and for profitability, and the bank manager will judge whether the business is likely to succeed. Thus the successful owner-occupier of many years' standing can borrow money to expand his business more easily than anyone hoping to start a business career, say as a tenant farmer, on very limited capital or without tangible security.

Bank Loans

Banks are able to provide a loan service. Long-term loans for the purchase of land or buildings are often called mortgages (see below). Shorter-term loans are also provided; in the farming context, such an arrangement is likely to be applied to cases where money is needed for developing a farm – what may be thought of as long-term tenant's capital. Starting a new pig enterprise, or building and equipping a new grain store can, subject to normal business criteria and creditworthiness, be financed in this way.

Rate of interest will be a few per cent above the base lending rate and there is usually a one-off fee charged at the outset for arranging the finance. The bank manager will expect to see a budget showing that the full sum is likely to be repaid within a predictable period of the order of five or seven years. The loan account is separate from the day-to-day current account. Payments into loan accounts are made by credit transfer from the current account, normally on a predetermined regular basis.

Mortgages

Banks may also provide mortgages for the purchase of land, farm houses and buildings. Generally it is a long-term arrangement; interest may be fixed or variable. Payments are made by direct debit, usually monthly, although agreement can be reached for other regular repayment timescales to match the cash-flow of the business. Once the mortgage is set up, there are no further annual renewal fees or reviews. Some banks include a clause to the effect that the mortgage is not automatically 'called-in' on the death of the borrower, so it can pass to the next generation.

Overdrafts

Many of the bank loan criteria apply to a bank overdraft, but an overdraft simply consists of allowing the current account to be overdrawn, with the permission of the manager, up (i.e., down) to a certain limit. Although some fairly major tenant's investments may be financed by overdraft, the traditional use of this facility is to provide seasonal credit while awaiting an inflow of funds later in the season. It may, for instance, fund the purchase of feedstuffs and payment of wages pending the time fat-stock can be sold. The bank manager will not normally expect the current account to be permanently 'in the red', although this depends on the length of the 'trading cycle', which can be very prolonged in some farming enterprises.

Interest is charged on the overdraft as on a loan account, but the overdraft is more flexible: interest is computed on a day-to-day basis so that every deposit made to reduce the overdraft immediately reduces the interest due. It can therefore be less expensive than having a loan account as well as a current account that runs in credit.

It is important, however, to remember that, whereas a loan facility cannot be called in unless the terms of the loan are not adhered to, an overdraft is 'repayable on demand'. In other words, the bank can remove the facility and demand the repayment of the borrowing without notice. It is very rare for this to happen without some warning procedures, so it is vital to keep the bank manager advised of progress, even when (or especially when) things are not going according to plan.

The term 'bank charge' covers all charges and fees made by a bank to their customers. These charges may take many forms, including:

- Monthly charges for the provision of an account.
- Charges for specific transactions (other than overdraft limit excesses).
- Interest in respect of overdrafts (whether authorised or unauthorised by the bank).
- Charges for exceeding authorised overdraft limits, or making payments (or attempting to make payments) where no authorised overdraft exists.

Other Sources of Credit

Farmers are well-known for being averse to change and traditionally use the same bankers for many years. This is likely to be the first port of call to discuss borrowing facilities. Other sources of money for farming include the following:

Merchant's Credit

Many suppliers of farm requisites will set up credit accounts for purchases after making a credit check of the business.

Hire Purchase and Leasing Schemes

There are a number of options that can be used instead of outright purchase of items of equipment, the following examples are the ones most commonly used in farming.

Hire purchase occurs when a finance company provides finance for the capital purchase chosen by the hirer. The hirer pays:

- The VAT on the capital sum, as part of the first payment (but it is reclaimable by registered businesses at the tax point).
- All insurance, running, maintenance, and repair costs.
- Charges for use of the finance company's capital, plus contributions to cover repayment of that capital, none of which involves VAT.
- At the end of the fixed term (of years, normally) the total paid will have covered the full capital cost and the equipment becomes the property of the hirer.

- **Equipment leasing finance** – a finance company (lessor) makes the capital purchase at the request of the farmer (the lessee). The lessee pays:
    - A charge for the use of the equipment and of the lessor's capital, by regular payments of fixed amounts; this is regarded as the provision of a service, so reclaimable VAT is added to each charge.
    - All insurance, running, maintenance and repair costs.
    - At the end of the fixed term (of years, normally), the total paid will have covered the full capital cost of the equipment but ownership is retained by the lessor.
    - Optionally, there may then follow a second period of fixed duration during which the regular payments are reduced to little more than an administrative charge.

Contract Hire

This is another way of acquiring new machinery.

- The machine is never owned by the farm but by the finance company.
- The machine is rented for a term ranging from one to seven years.
- The rentals are subject to VAT at the time of making the rental payment.
- A maintenance package is likely to be included in the agreement but it may not be.

- If a maintenance package is included, it is normally invoiced and paid separately at the time the machine is serviced.
- At the end of the term, the machine is returned and may be subject to returns conditions that are normally applied and agreed at the beginning of the agreement.
- An advantage of contract hire is the reduction in capital outlay.

All these choices have very significant implications in terms of financial planning, capital requirements, cash flow and balance sheet stability. They vary too in the degree to which the hirer or lessee maintains full management control of the resource in question. As such, they present decisions for the manager rather than the administrator.

Cash Transaction

However, cash may be important to businesses that include a retail enterprise such as a farm shop. Where an enterprise is cash-orientated, a till becomes a necessity. The till can be programmed to:

- Total the customer's purchases.
- Produce an itemised receipt for the customer.
- Allocate goods to the appropriate VAT rate.
- Allocate transactions to a chosen system of numbered categories.
- Scan goods using barcodes.
- Link to credit card terminals.
- Total the daily takings, with category subtotals.

Using the till printouts at the end of the day, it is possible to:

- Reconcile takings with receipts.
- Transfer figures to financial records.
- Check stock and re-order as necessary.

Petty cash refers to small amounts of cash kept on hand in a business. The two main reasons to keep petty cash:

- To accept cash for small sales at the farm gate.
- To pay for small purchases that require cash, such as an odd nut and bolt, postage, etc.

Petty cash should be securely stored in a cash-box and preferably in a locked safe or filing cabinet. The date, detail and amount of every transaction into and out of the box must be recorded and the amount in the cash-box should always reconcile with the petty cash account. The record, which may be in a notebook kept in the cash-box, is then transferred to the accounting system. Cash payments should always be supported with tax invoices wherever possible. Failing that, till tickets or other forms of receipts such as printed petty cash slips should be completed.

There are various methods of accounting for petty cash, some of which are listed below.

Out of Pocket System

This system is often used where a business has very few cash transactions. Where cash expenditure on business transactions is made by the owner over the past week or month, it can be repaid by cheque or online transfer. This repayment can then be entered into the accounting system, items analysed to the appropriate headings and input VAT reclaimed. This system has two advantages; no cash is kept in the office, and it is not necessary to run a separate petty cash account.

Float System

A suitable amount of money is drawn as cash from the bank. When this has nearly all been used, a similar amount is drawn again. Alternatively, if there are more receipts than expenses, the surplus cash is banked from time to time in the interests of security. The withdrawal or paying-in of cash can be recorded as a transfer from the current account to the petty cash account or vice versa, then payments are made from the petty cash with details coded or analysed in the same way as is done for non-cash transactions.

Imprest System

A suitable amount of money is drawn as cash from the bank. At the end of each month or other suitable period, the situation is checked and more cash drawn or some cash is deposited, bringing the petty cash balance back to the original sum. Such a system implies a more regular routine and ensures that checks are carried out regularly but is otherwise the same as a float system.

Large amounts of cash should be banked as soon as possible. Where the business is involved with a regular turnover of cash, it may be more cost-effective and secure to have the cash collected by a security company, particularly if it is a rural business with no banking facilities nearby.

Money Laundering Regulations 2007 require all accounting service providers (ASPs), including self-employed farm administrators, who are not supervised by an approved organisation, to register with HMRC. The regulations require bookkeepers to be sure of the identity of clients and to report any suspicions of money laundering to the National Crime Agency (NCA) by means of a SAR (Serious Activity Report). The regulations do not apply to employed farm administrators or family members who keep the farm accounts on a non-commercial basis. Full details can be found at www.gov.uk.

Top Tips

- Keep on top of your finances 24 hours a day with online banking.
- Make sure computer antivirus software is up-to-date, firewalls are enabled and browsers are set to the highest level of security.
- Choose a good password, at least eight characters with a mixture of upper- and lower-case letters, digits and symbols.
- Check your statements regularly, if you notice anything strange, contact your bank immediately.
- Cash collection services may be available for rural retail outlets.
- It is illegal to provide freelance accounting services unless registered under the Money Laundering Regulations 2007.

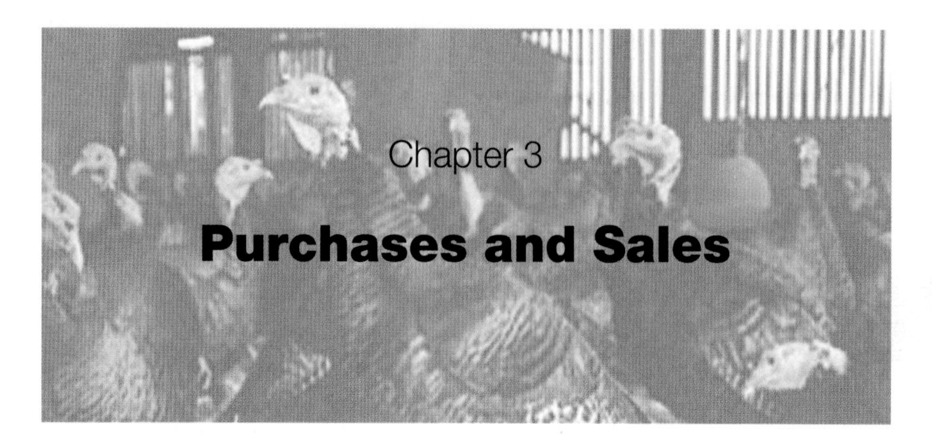

Chapter 3

Purchases and Sales

In this chapter we look at the document trail involved when buying goods and services in a typical rural business and selling the resulting produce.

It is possible to buy items from shops without the involvement of any paperwork at all. But business transactions virtually always require the safeguard of verifiable written evidence, for both practical and legal reasons. Use of the internet for managing purchases and banking is quick and efficient; use of a computer financial program for keeping track of money owed by the business and money owed to the business also makes good business sense.

Purchase Document Trail

Quotation

For the purchase of certain items and services, it is important to obtain comparative quotations – written statements setting out the cost for a particular job or service – and these should always be obtained before a large order is placed. A quotation may be an estimate and may cover aspects of the purchase other than cost, such as packaging and delivery details.

Purchase Order

For general purchases it is preferable to have a written order: emails are acceptable by many suppliers. When orders are placed via a supplier's website, confirmation may be automatic and immediate. Alternatively, a duplicate book can be used: the top copy is sent to the supplier and the duplicate retained. Where a business has many suppliers – for example, a farm shop, a computerised purchase order processing module may be added to the accounting system.

Sales Contract (Issued by Merchant)

Fertiliser and other commodities may be purchased on a contract where it may be advantageous to fix a delivery period and price. Details should be retained and tied in with the purchase invoice.

Order Confirmation

A supplier will generally confirm the arrival of an order, unless it is a routine matter and likely to be dealt with at an early date. It may take the form of a document or an email, which should probably be printed for security reasons.

- It may simply list the items ordered but will allocate a serial or job number to which reference can be made if necessary.
- It may substitute supplier's reference numbers in respect of the items ordered. It is not unknown for suppliers or their computers to substitute 'nearest available' in place of out-of-stock items: beware!
- Predicted delivery details may be specified, particularly for capital items and especially if being made to order.

The amount of detail varies. It is not unknown for the confirmation to take the form of a fully VAT-compliant invoice: indeed, that is a requirement in cases of cash with order.

Despatch Note

If an item is sent by courier, the supplier may send a despatch note. This gives details of the goods, method of transport, despatch date and instructions of who to contact in the event of non-delivery. There may also be details of action to be taken in the event of the goods being damaged in transit. It can be 'filed' in the diary at the predicted delivery date.

Collection Note

When goods are purchased from a local supplier and paid for on account, the person collecting the goods will sign a collection note, which can then be matched with the purchase invoice when it is received.

Delivery Note

A delivery note will arrive with the goods; a signature may be required by the driver as evidence of the goods being complete and in good condition. Delivery notes relating to certain farm inputs such as seed, fertiliser and chemicals may contain important information not included on the purchase invoice and must be retained for compliance evidence. See Chapter 12.

Order Discrepancies

The goods should be checked against the order and any discrepancies should be taken up with the supplier promptly.

Matching Delivery Notes with the Purchase Invoice

It is important that a reliable method of collecting and retaining documents relating to farm deliveries is in place: details will depend on the size and nature of the business. However, in every farming situation, all farm staff should be made aware of the importance of making sure that delivery/collection notes reach the farm office where they can be tied in with the purchase invoice.

Purchase Invoice

A purchase invoice is entered in the books of the business buying the goods. (This same invoice is the sales invoice of the seller of the goods). It should be checked against all other documentation relating to the purchase. Any discrepancies must be referred to the supplier. VAT regulations prohibit the alteration of invoices. The purchase invoice is a legal document and must be retained with the annual accounts and records for seven years. The following example shows the details required on a purchase invoice (and a standard sales invoice).

Purchase Credit Notes

A credit note may be raised where goods are returned or to rectify a mistake made on a purchase invoice. It is processed in the same way as invoices and should never be recorded as a receipt.

Processing Purchase Invoices and Credit Notes

Once the invoice or credit note has passed the various checks mentioned above, it should be authorised for payment by the business principal or

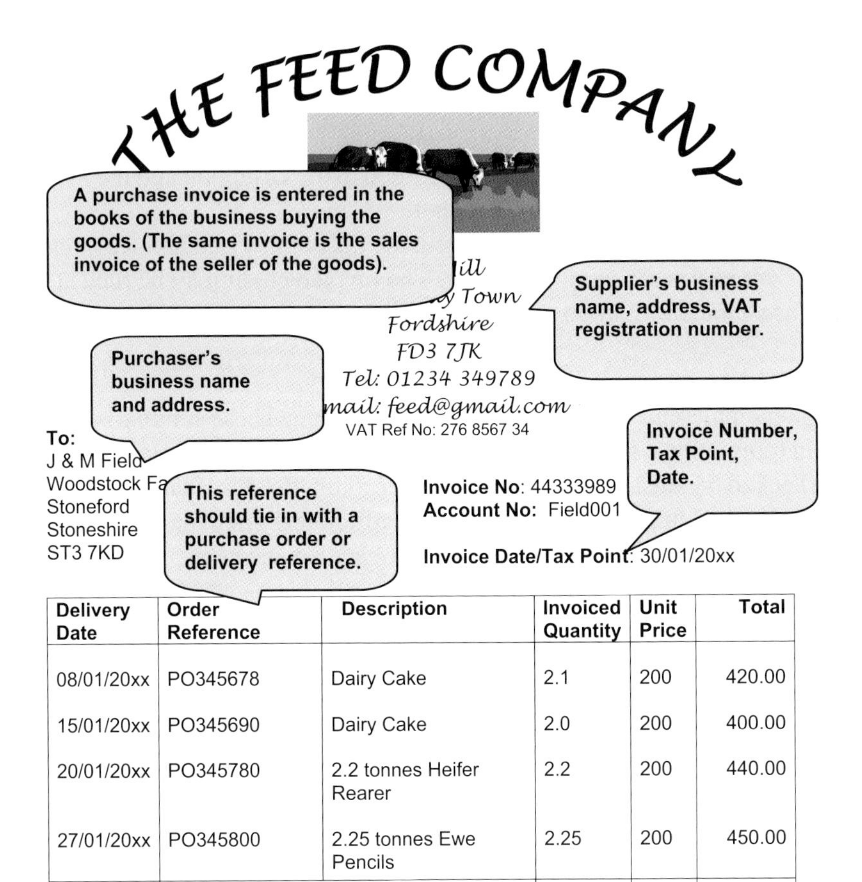

THE FEED COMPANY

ill
y Town
Fordshire
FD3 7JK
Tel: 01234 349789
mail: feed@gmail.com
VAT Ref No: 276 8567 34

To:
J & M Field
Woodstock Fa
Stoneford
Stoneshire
ST3 7KD

Invoice No: 44333989
Account No: Field001

Invoice Date/Tax Point: 30/01/20xx

Delivery Date	Order Reference	Description	Invoiced Quantity	Unit Price	Total
08/01/20xx	PO345678	Dairy Cake	2.1	200	420.00
15/01/20xx	PO345690	Dairy Cake	2.0	200	400.00
20/01/20xx	PO345780	2.2 tonnes Heifer Rearer	2.2	200	440.00
27/01/20xx	PO345800	2.25 tonnes Ewe Pencils	2.25	200	450.00
Sub total					**£1710.00**
<u>**VAT Code**</u> Zero	<u>**Rate**</u> 0%				**0.00**
Total Net Due					**£1710.00**

Payment Terms: **30 days from date of invoice**
Bank Details: Sort Code: 12:13:14 Account No: 01235890

Company Registration No 5290123
Registered Office: Unit 3, Faraway Business Park, Fordshire, FD3 7JK

Figure 3.1 A purchase invoice

suitably designated person. Where a cashbook system is in place, the invoice may simply be paid and any credit notes deducted, according to the terms, or a payment may be requested in respect of the credit where there is no invoice due. The date and cheque number are noted on the invoice, the cheque despatched, the transaction recorded in the cashbook and the invoice and any credit note filed. See Chapter 1.

Where a double-entry system is used, the duly authorised invoice or credit note is entered in the purchase ledger. The invoice/credit may be filed at this stage as all the information needed to make the payment is now on the system.

Statements

Suppliers may send monthly statements of account. These summarise all the recent transactions, finishing with a balance owing to or by them; each should be checked against the corresponding account details. Statements alone cannot be used in place of invoices and credit notes. They may be filed with other account documents, or used as remittance advice notes, or discarded.

Purchase Payments

The routine for making payments will depend on the system in place for the individual business. Many computerised accounting packages have the facility to export payments to banking software. This saves considerable time and expenditure on postage and bank charges.

Remittance Advice

Many computer accounts programs will produce a remittance advice and may also incorporate emailing this to the supplier. Some statements/ invoices even have tear-off pre-printed remittance advice sections. Where nothing else is available, a compliments slip may be used to accompany a cheque, with a note of the invoice number.

Sales Document Trail

Purchase Contracts (Issued by Merchant)

Grain and other commodities are generally sold on contract with an agreed delivery period and price. These should be retained and tied in with deliveries against the contract.

Grower Contracts

Some agricultural products such as sugar beet and other specialised crops are grown on contract with the processor. These are important documents but not generally referred to with regard to accounts.

Log Book of Goods Leaving the Farm

A record of all goods leaving the farm should be kept in a log book or at least a day book/diary. The record should include the date, the detail of goods, the approximate weight of load, the number of animals, the destination and name of haulier. Collection/delivery documentation and subsequent self-billing invoices can be checked against this record.

Collection/Delivery Documentation

Many farm outputs are covered by self-billed invoices provided by purchasers. Farm gate sales may be recorded through a cash system. Where a business is involved with adding value to farm produce and numerous sales are generated, the nature and sequence of documents will be similar to those described above under purchasing. It may be advantageous to add a sales order processing module to the accounting system.

Where farm produce is concerned, the producer will usually be given instructions by the purchaser on the collection and documentation required. Milk is one of the few commodities that is measured accurately before it leaves the farm, but only the quantity; quality is measured in the buyer's laboratories from a sample taken by the driver. The tanker driver will produce a printout showing litres collected. The litreage should be recorded in the milk book, which can be checked against the monthly milk statement and act as a milk quota record.

For other bulk commodities such as grain and potatoes collected by lorry, which holds approximately 29 tonnes, collection tickets will be left by the drivers. These should be kept and tied in with subsequent confirmation of intake weights. It is also important to keep a running total of loads against the purchase contract in order to ascertain when a contract is complete.

The process for collecting sugar beet is automated and the haulage contractor will be issued with a delivery permit for each load (approximately 29 tonnes), which will be logged into the computerised system at the factory. In due course, the permits will be returned to the grower with the haulage invoice. It is not usual for the driver to leave a collection note but it may be the farm policy and good practice to get the driver's signature in the log book of goods leaving the farm.

Additional paperwork may be required by law to accompany the goods when they leave the farm, particularly for livestock. If the goods are produced to a quality standard such as organic, a copy of a current certificate of compliance will also be required. More information and templates of movement documents can be found at www.nfu.co.uk or www.gov.uk. See also Chapter 12.

Sales and Self-Billing Invoices

The number of invoices raised will depend on the nature of the business. In a family farming situation with little income other than for farm produce, few sales invoices will be issued. Generally, the quantity and quality of the goods that leave the farm are not accurately measured until they are collected by, or delivered to, the buyer who will issue a self-billed invoice based on this information and any purchase contract which may be in place. Following is an example of a monthly invoice for the milk collected during the month. A summary of each collection will accompany the invoice which may be checked against collection tickets.

A standard sales invoice will contain the same basic information as shown in Figure 3.1. However, the self-billing invoice, Figure 3.2, may be more complex and will contain other useful physical information besides financial, such as tonnes of grain, numbers and quality grading of livestock. A self-billing invoice will also show deductions, sometimes referred to as contras. These may include levies, weighbridge charges, haulage charges and marketing commission plus VAT where applicable, which is reclaimable. Occasionally, a farming business may need to raise a self-billing invoice, but before this may be done, an agreement must be in place with the supplier. An example of an agreement plus the rules covering the self-billing process can be found at www.gov.uk.

Sales Credits

A sales credit note is the converse of a purchase credit as described above.

The Milk Company

The Dairy
Faraway Town
Fordshire
FD3 7JK
Tel: 01234 349789
email: milk@gmail.com
VAT Reg. No: 276 8567 34

Member No: 250801333
Supplier's Name: J & M Field
Supplier's RPA No: B26587M
Supplier's VAT No: 156 7321 54

Seller's business name and address.

Su

J & M Field
Woodstock Farm
Stoneford
Stoneshire
ST3 7KD

SELF BILLING INVOICE

Issued by the buyer of the goods.

Invoice No: 60001234

Invoice Date/Tax Point : 05/02/20xx

Self Billing Invoice / Statement for January 20xx

Your Milk Price	Litres	PPL	Amount £
Butterfat	81204	0.09014	7319.72
Protein	81204	0.1323	10744.91
Bactoscan	81204	000	
Volume Bonus	81204	0.0022	178.65
Factory Bonus	81204	0.003	243.61
			18486.89
		Net PPL (total milk value/total litre)	22.766
		Net PPL – 12 mths Rolling Average	23.012

The quantity and quality of the goods that leave the farm are not accurately measured until they are collected by or delivered to the buyer. Collection details/references may be shown in a separate schedule and summarised on the self-bill.

VAT Rate	Your Invoice to Us	Litres	Amount £
ZR	Milk	81204	18486.89
			18486.89

VAT Rate	Our Invoice to You	Amount £
01	AHDB Levy	55.41
S1	VAT Code 6 (20.00%)	11.08
		66.49

The VAT shown above is your input tax reclaimable.

Where more than one VAT rate is used, the total net value for each should be shown on the invoice.

Balance Brought Forward	**0.00**	
Your Invoice to Us	**18486.89**	
Our Invoice to You	**-66.49**	
Amount Due for Payment to your bank account on 10/02/20xx		**18420.40**

Company Registration No 5290123
Registered Office: Unit 3, Faraway Business Park, Fordshire, FD3 7JK

Figure 3.2 A self-billing invoice

Processing Sales Invoices and Sales Credit Notes
Once the sales invoice or credit note has been raised, a system needs to be in place to keep track of money owed to the business. With a cashbook system, this may be a simple list of sales invoices with a column in which to record the date when an invoice has been paid. Where a computerised double-entry system is used, the amount owed to the business is recorded on the customer's account in the sales ledger, where it will be shown as outstanding until payment has been received and recorded.

Sales Receipts
Many farm and horticultural businesses sell their major products to, or through, organisations that make direct credit payments as set out in the self-billed invoices that those organisations have originated; those invoices may well serve also as remittance advice notes. Any cheques should be banked promptly and all receipts recorded in the cashbook (single entry) or in the customer sales ledger (double entry). Not all receipts will arrive promptly; the list of sales invoices issued should be monitored regularly in order to chase late payments by sending a statement or reminder or in some cases with a phone call.

Cooperatives
There are many benefits for a farming business in joining a cooperative group of similar businesses, buying or selling similar goods and services. Cooperative organisations include specialised trading groups, selling milk or grain, for example, or packing and wholesaling horticultural produce; more general purchasing groups buying crop or livestock inputs; and machinery 'rings'. Regional farmer member groups are growing, offering both trading and purchasing services.

Cooperative Trading and Marketing Groups

The farm administrator will also see benefits from the use of cooperatives, particularly their online facilities. If produce – be it vegetables, grain or milk – is all sold to a cooperative, the paper trail will be computerised and uniform. Soon after a load of grain has been weighed in by the cooperative and checks on moisture have been made, the results can be viewed online.

Similarly with milk or flowers, physical and financial data are available online.

When goods leave the farm, a milk ticket or a collection note will be generated by the lorry driver, thus starting the administrative process for the particular cooperative. It may be a year before self-billing invoices for grain sold to a cooperative are received. In the meantime, payments on account may have been drawn down; these will be shown as deductions on the final self-billing invoice, which will also show any remaining balance due.

Cooperative Buying Groups

When using a buying group, purchase invoices from a variety of suppliers will be raised at point of sale then sent to the buying group, which checks and batches together the invoices and credit notes generated that month. They are then attached to a monthly statement, itemising each transaction and sent to the farm for settlement, which is usually by monthly direct debit.

Purchase invoices/credits still need to be checked against delivery/return notes. The invoice/credit is raised by the supplier on the account of the buying group and will show the farmer member's account number.

Additional benefits are the savings to be made in time spent on the telephone and processing multiple supplier payments, bank charges and postage. The larger cooperatives use modern technology and many offer useful online facilities to members. On the purchasing side, purchase invoices/credits are available to view or download from the member's online account; this facility is useful if needing to refer to an invoice or credit note before the end of the month when the hard copy arrives.

Each transaction included on the statement needs analysing. This may be done manually for a cashbook system, totalling the VAT and any items with a common account heading. The treatment of levies (charges made by the group) to the members varies from group to group. In some cases, they are shown as deductions on the statement each month and analysed according to the type of purchase. VAT is charged on the levy and may be reclaimed. Other groups send a separate invoice for a six-month period showing the turnover for the period of each type of purchase. In this case, it may be necessary to apportion the net cost to the relevant account heading on a percentage basis.

Where a double-entry computer system is used, a purchase ledger account in the name of the buying group is created and transactions with many different suppliers are managed in one account.

From the farm administrator's point of view, whatever the cooperative, there will be a sound paper trail in place, with guidelines on the process and help just a phone call or an email away.

Top Tips

- Encourage staff to hand in delivery notes promptly with the addition of useful information such as which machine or building the delivery refers to.
- Use a stack of three or four trays and sort mail as it arrives.
- Use customer and supplier names as account references, rather than numbers.
- Tie in self-billing invoices for grain etc. in with purchase contracts and keep a running total of tonnages delivered against the contract.
- Use customer and supplier online facilities to view accounts and download invoices and other reports.

Chapter 4

Single-Entry Cash Analysis

Financial records of any business must be kept by law for six years, but the law does not prescribe the method of recording transactions. In this chapter, we look at a single-entry cash analysis system suitable for the accounting needs of a small business. This system has its limitations and it is important to consider how this simple system will best serve the needs of the business. The primary functions are to enable the VAT return to be completed and to facilitate the production of the profit and loss account and balance sheet – necessarily distinguishing revenue from capital, business from private, and so on. In a business with many customers and suppliers, a double-entry system is recommended, see Chapter 6. For management accounts such as budgets and gross margins, further analysis is required to take account of production cycles and a separate recording system is recommended. See Chapter 11.

While a manual cashbook, (often referred to as a ledger) can be kept, it is now far more practical and efficient to use a simple computerised spreadsheet program to maintain a cash analysis system. Keeping 'the books' up-to-date is key, at the very least monthly, when the bank statement arrives and also ahead of the VAT return. Another option is to use a single-entry accounting program designed for small businesses, with ease of use rather than complex accounting functions in mind. Contact details of specialist software providers can be found in Appendix 1.

We have chosen a VAT registered farming business model for examples etc., but the principles and accounting rules are similar for any small VAT registered business. The cashbook examples are based on a spreadsheet with automated totals shown above the analysis columns (illustrated on the spreadsheet at the end of the chapter). A paper-based system can be designed along the same lines with totals shown at the bottom of the page.

Cash Accounting

For the recording of VAT, this system is referred to as 'cash accounting', where VAT is accounted for on the date of bank transaction. HMRC rules on the cash accounting scheme for VAT can be found at www.gov.uk. The alternative system for recording VAT is 'invoice basis', where the VAT is accounted for on the tax point date of the invoice rather than the date of the bank transaction. For this method a double-entry system is strongly recommended rather than trying to adapt a cash accounting system.

Setting up a Cash Analysis System: Expenditure Column Headings

Date	Detail	Ref	Cleared	Amount	Contra	Net Amount VAT Inputs	Outside Scope of VAT	VAT to reclaim on purchases	Variable Costs	Overhead Costs	Other Costs	VAT Payable to HMRC

Diagram 4.1

Date & Detail	**Reference**	**Cleared**	**Amount**
of payment and supplier	Cheque number SO or DD	Bank check column	Total of transaction

- **Contra column:** This column is used to record a receipt that has been offset against a purchase, or vice-versa, as explained below.
- **VAT headings:** This is the group of headings used to record the information necessary to complete the VAT return.
- **Net amount (VAT inputs):** Record all standard, zero and exempt inputs (purchases, etc.) net of VAT – the total (for the quarter or month) will be entered in Box 7 on the VAT return.
- **Outside scope of VAT:** Record all items outside the scope of VAT. Examples include wages, taxes and private drawings. The total is used to balance figures and is not included on the return.

- **VAT to reclaim on purchases:** Amount of VAT to be reclaimed from HMRC on inputs, the total (for the quarter or month) will be entered on the VAT return in Box 4.

Variable Cost Headings

These are items of trading expenditure, specific to an enterprise, and which vary in direct proportion to the size of any proposed change in an enterprise (also referred to as direct costs).

The choice of headings will depend upon enterprises; some examples for a farm are:

Concentrates
Bulk feeds
Veterinary services and medicines (*Vet & med*)
Artificial insemination
Replacement livestock purchases
Livestock sundries
Contract labour and services
Seed, plants and bulbs
Fertiliser
Crop protection
Crop sundries
Contract/hire/haulage (*where enterprise-proportional*)
Miscellaneous variable costs

Overhead Costs Headings

Costs that are incurred in the general running of the business and are not usually attributable to any particular enterprise, (also referred to as fixed costs):

Wages and PAYE/NI
General contract (*e.g., hedging*)
Liming (*not included in VCs as seldom benefits single crop*)
Machinery repairs, spares and insurance
Fuel and fuel scale charge
Vehicle repairs/tax/insurance
Property repairs and consumables

Office/telephone/professional fees
Electricity
Rent/rates/water/general drainage charge
Bank charges/interest *(including interest element of mortgage or HP repayments)*/finance charges
Other fixed costs
General insurance

Other Headings
Capital (outright purchase of assets and capital element of Mortgage or HP repayments)

Private drawings
Income tax
VAT paid to HMRC

Accounting Headings
These are optional, not shown in Diagram 4.1:

To Deposit Account
To Loan Account
To Petty Cash

Income and Rcceipts Column Headings

Date	Detail	Ref	Cleared	Amount	To Bank	Contra	Net Amount VAT Outputs	Outside Scope of VAT	VAT charged on sales	Enterprise Income	Other Income	VAT Payable to HMRC

Diagram 4.2

Date & Detail	**Reference**	**Cleared**	**Amount**
of receipt and customer	Counterfoil no., direct credit	Bank check column	Total of transaction

- **Contra column:** This column is used to record an expense which has been offset against a receipt or vice versa, as explained below.

- **VAT headings:** The group of headings used to record the information necessary to complete the VAT return.
- **Net amount (VAT outputs):** Record all standard, zero and exempt outputs (sales/receipts) net of VAT, the total (for the quarter or month) is entered in Box 6 of the Vat return.
- **Outside scope of VAT:** Record all items outside the scope of VAT. Examples include insurance settlements, intakes of fresh capital, and any private receipts. The total is used to cross-cast and balance the columns with the bank and contra columns and is not included on the return.
- **VAT chargeable on outputs** (sales), the total (for the quarter or month) will be entered on the VAT return in Boxes 1 and 3.

Enterprise Income (Outputs)	**General Income**	**Other Headings**
Milk, calves, cull cows	Rents	Government support and grants
Lambs, cull sheep, wool	Contract work	Capital sales
Wheat, barley, straw	Miscellaneous sales	Private income
Sugar beet	Interest earned	VAT refund from HMRC
	Sundry farm Income	

Accounting Headings

These are optional, not shown in Diagram 4.2.

From Deposit Account
From Loan Account
From Petty Cash

Cash Analysis – Entry Process

To complete the monthly cash accounting records and reconcile the bank account the following items will be required.

- **Cheque book stubs:** All cheque books in current use will be required in order to enter transactions.
- **Bank paying-in books:** Number and date as for cheque book stubs above.
- **Bank statements:** Ensure that bank statements are received at the

end of each month. An online banking facility is helpful as statements are always accessible.

- **Documentation:** As required by law supporting each transaction, for example:
  - Purchase invoices, sales invoices and self-billed invoices. See document trails in Chapter 3.
  - Mortgage schedule, hire purchase or leasing schedules, licence renewals, membership subscriptions, etc.
- **Other transactions:** Such evidence as is available in respect of cash and credit/debit card purchases including VAT wherever appropriate and possible.

Further Guidance on More Complex Transactions

The majority of entries in the accounts will be simply a matter of entering the date, detail, reference, bank amount, VAT value and an entry in the appropriate analysis column/s. However, there are various common anomalies that need further explanation.

Expenditure and Other Payments

Use of the contra columns (transaction which is a net expense): A machinery trade-in is a good example. The amount paid is entered in the bank column and the total amount received for the trade in is entered in the contra column. The net value of the new machine is entered in the relevant analysis column (capital expenditure), VAT in VAT to reclaim and the net value in Net amount VAT inputs. Then the recording of this transaction is completed on the receipts side – with the total received for the old machine entered in the contra column, the net value of the machine in the relevant analysis column (capital receipts), VAT charged on Sales and net amount VAT outputs. Contra columns on payments and receipts must always be identical. See diagram 4.3 on the next page.

Multi invoices: Purchases made via a buying group (cooperative) need pre-analysis before recording in the cashbook. See diagram 4.4.

Private apportionment: Small businesses and particularly farming businesses are often operated from home; private apportionment refers to the split in expense between private and business.

	February 20xx	Expenditure			VAT Recording Headings			Analysis Headings		
Date	Detail	Ref	Bank Amount	Contra	Net Amount VAT Inputs	Outside Scope of VAT	VAT to reclaim on purchases	Capital Purchase	Private	Comment
12/02/**	Burgett	112	360.00	120.00	400.00		80.00	400.00		New brushcutter less old Trade-in

	February 20xx	Income			VAT Recording Headings			Analysis Headings		
Date	Detail	Ref	Bank Amount	Contra	Net Amount VAT Outputs	Outside Scope of VAT	Vat charged on sales	Capital Sales	Private Income	Comment
12/02/**	Burgett			120.00	100.00		20.00	100.00		Chain Saw Trade-in

Diagram 4.3

	February 20xx	Expenditure			VAT Recording Headings			Analysis Headings			
Date	Detail	Ref	Amount	Contra	Net Amount VAT Inputs	Outside Scope of VAT	VAT to reclaim on purchases	Feed	Property Rprs & Consmbls.	Private	Comment
12/02/**	Farmers Co-op	110	787.72		761.65	23.24	2.83	747.50	14.15	23.24	Bulk feed – Dairy Cows

Invoice Detail

Description	Quant	Unit Price	Net Total	VAT Rate	VAT	Invoice Total
Bulk Feed	5.75	130.00	747.50	Z	0.00	747.50
Light Bulbs	1	3.90	3.90	Stnd	0.78	4.68
Wellington Boots	1	10.25	10.25	Stnd	2.05	12.30
Horse Sundries (private)	*1*	*19.37*	*19.37*	*Stnd*	*3.87*	*23.24*
Total payable					6.70	787.72

Analysis information:

Bulk feed – Dairy Cow feed
Light bulbs – Property repairs
Wellingtons – Consumables
Balance of items – treat as private and don't claim the input VAT.

Diagram 4.4

	February 20xx	Expenditure				VAT Recording Headings			Analysis Headings			
Date	Detail	Ref	Cleared	Amount	Contra	Net Amount VAT Inputs	Outside Scope of VAT	VAT to reclaim on purchases	Office & Prof Fees	Insurance	Private	Comment
10/02/**	TelCo	107	*	208.10		115.61	69.37	23.12	115.61		69.37	2/3rds business – 1/3rd private split agreed with HMRC

Diagram 4.5

There are no standard apportionments of private elements of expenses paid by a business, it is a matter for the business principal (or the accountant) to agree with HMRC fair and reasonable apportionments for accounting and VAT purposes. The following example is for a telephone bill with a split agreed with HMRC of 2/3 business and 1/3 private.

House improvement/repairs: If the cost relates to the farmhouse that has a split use (i.e., living and office elements) ask the following questions:

- Does the cost relate purely to the private parts of the house, e.g., bedroom? If so, this is defined as 100% private use.
- Does the cost relate purely to the business part of the house, e.g., office? If so, this is defined as a 100% business expense.

Otherwise, split expenditure based on agreed business to private proportions need to be applied. VAT does not necessarily follow accounting apportionments. See Chapter 7.

Water, electricity and gas: To complicate matters further, the bill may refer to more than one property, which also needs to be taken into consideration.

Fuel: It is common practice to analyse all fuel in one column, but it is important to identify domestic heating oil so that this can be apportioned.

- Gas oil (red/tractor diesel).
- Derv (white/vehicle diesel, possibly including private use).
- Kerosene/burning oil – for domestic heating and business oil usage.

Repairs to vehicles: Identify the vehicle the expense relates to. Generally VAT can be reclaimed in full for vehicle maintenance and repairs as long there is some element of business use.

Private expenditure: The only fully recognised evidential documents in the normal course of business are invoices and payment schedules. Shop purchases may be verified with 'simplified' VAT invoices in the form of till tickets specifying the VAT number etc. in which case, value and VAT can be calculated from the price using 'the VAT fraction'. A cash-only till ticket may be acceptable as a business expense but no VAT may be claimed. In the absence of any significant evidence such small purchases must be regarded as 'Private'.

Invoices inclusive of VAT: Where an invoice total is inclusive of VAT, use the *VAT fraction* for the current rate. See Chapter 7.

Credit charges and discounts: Such charges and discounts are always calculated on the net amount before VAT.

Refunds: If goods are returned or prepaid services terminated, an actual refund may result. (Other credit notes can simply be deducted next time round.) The refund of an earlier expense must not be treated as a receipt: it must again be entered among expenses but as negative amounts right across the line including VAT columns. (Spreadsheets can normally be set so that negative amounts appear in red.)

Capital versus repairs: Where significant works are being carried out and there is an obvious improvement rather than a repair this should be analysed under capital spend; for example, where a gravel road is resurfaced with tarmac. Any significant works should be discussed with the farmer and accountant: capital works receive significantly different treatment in the annual (tax) accounts as compared with repairs.

Hire purchase: Diagram 4.6 shows how the agreement may be recorded over a number of months. The first payment contains all of the VAT therefore a manual adjustment needs to made for the net value of the asset to be recorded in Box 7 on the Vat return correctly. This is a somewhat anomalous situation, in that there is no normal relationship between the value of that initial transaction and the VAT being charged. There is also the interest element (sometimes referred to as finance charge) to be recorded. The HP repayments as shown in

Expenditure				VAT Recording Headings			Analysis Headings		
Date	Detail	REF	Amount	Net Amount VAT Inputs	Outside Scope of VAT	VAT to reclaim	Bank Charges, Interest & Finance Charges	Capital	Comment
15/02/**	FG Equipment Finance	118	3077.34	*1000.00		2077.34		1000.00	Bailey Trailer
Entries for the following 36 months:-									
15/03/**	FG Equipment Finance	DD	298.00		298.00		37.42	260.58	
15/04/**	FG Equipment Finance	DD	298.00		298.00		37.42	260.58	
15/05/**	FG Equipment Finance	DD	298.00		298.00		37.42	260.58	
15/06/**	FG Equipment Finance	DD	298.00		298.00		37.42	260.58	
15/06/**	FG Equipment Finance	DD	298.00		298.00		37.42	260.58	
30 × the same monthly entry		DD	8940.00		8940.00		1122.74	7817.40	
15/01/**	FG Equipment Finance	DD	304.00		298.00		37.45	266.55	
			13811.34	1000.00	10728.00	2077.34	1347.29	9386.71	
*Manual adjustment required for Box 7 on VAT Return				9386.71					
				10386.71					

Hire Purchase Agreement Detail	
Duration	36 months
Amount of credit	£9386.71
Total payable	£13811.34
Interest charged	£1347.29
35 monthly instalments	£298
1 final instalment	£304

Invoice Detail	
Net cost of goods	£10386.71
VAT @ standard rate	2077.34
Total payable	**£12,464.05**

Diagram 4.6

the schedule should be split between capital and interest, as shown in the diagram. The data to be recorded in the account must agree with the details of the HP Agreement.

Leasing or rental payments: An invoice and schedule is raised by the leasing or rental company showing the VAT that may be reclaimed. The net value of the invoice is then analysed as an overhead under 'lease/rental payments' or under MRS (Machinery Repairs & Spares, etc.) in the case of a machine.

Capital purchases: Full details need to be recorded (comments column) of capital purchases, in particular, the clear identification of the item in relation to its make/registration number.

Taxation: Income Tax and NIC relating to partners should be recorded as private drawings but PAYE & NIC relating to employees (including directors) are recorded under Wages & PAYE.

Payment on account: Where an invoice is not paid in full, VAT should only be reclaimed in proportion to the amount paid on account at the time of payment. The balance can be reclaimed as subsequent payments are entered in the cashbook. Photocopy such invoices, with annotations, in support of each payment and VAT claim.

Petty cash: A simple way of dealing with petty cash transactions is to recompense the monthly total of the cash transactions with a cheque (or to bank the surplus) which can then be analysed in the cashbook. See Chapter 2.

Fuel scale charge: Fuel scale charge is explained in Chapter 7. Diagram 4.7 demonstrates how the VAT element is recorded in the cashbook. Note there is no cash movement.

	February 20xx	Income		VAT Recording Headings					
Date	Detail	Amount	Contra	Net Amount VAT Outputs	Outside Scope of VAT	Vat charged on sales	Capital Purchase	Private Income	Fuel Scale Charge
28/02/**	Fuel Scale Charge		80.00	66.67		13.33			66.67

	February 20xx	Expenditure		VAT Recording Headings					
Date	Detail	Amount	Contra	Net Amount VAT Inputs	Outside Scope of VAT	VAT to reclaim on purchases	Capital	Private	Fuel Scale Charge
28/02/**	Fuel Scale Charge		80.00		80.00	0.00			80.00

Diagram 4.7

Income and Receipts

Self-Billed Invoice (including contra transaction, which is a net receipt)

A self-billed invoice is generated by the purchaser of the produce; Diagram 4.8 shows 26 fat lambs sold at market. Levies have been deducted from the sale proceeds and this element of the transaction is recorded using the contra column. The amount received is entered in the bank column and the total amount of expenditure deducted in the contra column. The net value (Zero rated in this example) of the produce is entered in the relevant VAT and analysis columns). The recording of this transaction is completed with the same amount (as receipts contra column) entered in the contra column on the expenditure side, the net to the relevant analysis column (expenditure) and VAT where applicable in VAT to reclaim and net amount VAT inputs. Contra columns must always balance.

	February 20xx	Income			VAT Recording Headings			Analysis Headings			
Date	Detail	Amount	Cleared	Contra	Net Amount VAT Outputs	Outside Scope of VAT	Vat charged on sales	Fat Lambs	Cull Sheep	Wool	Comment
08/02/**	Faraway Market	1904.00	*	96.00	2000.00		0.00	2000.00			26 Lambs

	February 20xx	Expenditure			VAT Recording Headings			Analysis Headings			
Date	Details	Amount	Cleared	Contra	Net Amount VAT Inputs	Outside Scope of VAT	VAT to reclaim on purchases	Feed	Livestock Purchase	Stock Sundries	Comment
08/02/**	Faraway Market		*	96.00	80.00		16.00			80.00	

Diagram 4.8

Private Income and Capital Introduced

Introduction of private funds must be clearly identified and should be analysed under private income. This may be useful for an accountant when preparing personal tax returns and provides a clear audit trail.

Government Support Payments and Grants received from DEFRA (Rural Payments Agency)

The treatment of this income for tax purposes depends on the type of grant or individual scheme and it is therefore important that all such receipts are clearly identified by source and scheme-end year.

Comments Column (Expenditure and Income)

The 'Comments' column may be used to record other useful information, for example:

- **Physical quantities:** Include numbers of animals/tonnes of grain in the comment column. An important aspect of recording is the ability to 'test' the completeness and accuracy of the figures provided against the physical output of an enterprise. For example, an accountant will compare numbers of animals sold (as per self-billed invoices) with a livestock reconciliation at the year-end. The same applies to tonnes of grain. Conversely tonnes of seed or feed purchased may be compared to the area grown. Comments boxes or an extra column can be used to record this information in a spreadsheet or extra lines can be used in a manual system. A cash accounting software program will probably have a facility for recording such physical information. Having the information to hand in the cashbook saves time.
- **Harvest years:** a key accounting principle is to match incomes to expenses incurred in that production cycle. It is therefore important that the harvest year is identified.
- **Insurance claims:** It is important to identify the reason for a particular insurance refund in order to be able to record the refunded amount against the original cost centre; for example:
    - Damage to agricultural buildings to be identified to building and property repairs.
    - Damage to vehicles to be identified against machinery repairs.
    - Refund of premiums if an asset has been sold.
    - Asset claim to compensate for a written off vehicle or property.

  Such insurance receipts are 'outside scope' of VAT.

- **Capital purchases and sales:** Full details need to be recorded of capital transactions, in particular, make/model and registration/serial numbers of agricultural machines and vehicles. The asset register will also need to be updated. This is dealt with in Chapter 8.

Checking the Totals in the Cashbook

Having entered all information from cheque-book stubs and paying-in books, work through the bank statement and enter direct debits, standing orders and BACS payments. If all information has been analysed correctly, the totals of the analysis columns should equal the bank total plus the contra column and the next step is to complete the Bank Reconciliation.

Bank Reconciliation

This calculation checks the accuracy of the cashbook against the bank statement. Where an accounts package or spreadsheet is used the reconciliation procedure is the same, the only difference is the automation of calculations.

Checking the Bank Statement Against the Cashbook

Working from the bank statement, tick each receipt that appears on the statement (cleared the bank) then find the same entry in the cashbook and put a * in the cleared column. Then work through all the payments in the same way.

Items in the cashbook, which have not appeared on the bank statement, are known as 'unpresented'; in a manual system they can be marked o/s (Outstanding).

Completing the Reconciliation

- Unless starting from a nil balance on a new bank account there will be a closing balance for the previous period, which becomes the opening balance in the reconciliation.
- Unpresented cheques – total of items not 'ticked' on the expenses sheet.
- Unpresented income – total of items not 'ticked' on the income sheet.

Example of a Bank Reconciliation Calculation

Bank Statement

a	Bank statement balance	–1999.89
b	– Unpresented cheques	–5234.64
c	+ Unpresented income	+0.00
	= Closing balance at end of the month	–7234.53
d	Cashbook balance bought forward from the previous period	
e	Total from bank column (expenses)	
f	Total from bank column (receipts)	

Cashbook

d	Cashbook balance	–4832.00
e	Expenses as per bank column	–23710.45
f	Income as per bank column	+21307.92
	= Closing cashbook balance	–7234.53

The closing bank balance (adjusted for unpresented items) should reconcile (agree) with the closing cashbook balance.

How to Reconcile the Cashbook against the Bank Statement Where an Opening Balance is Not Available

If a set of the previous year's accounts are available, the closing bank balance (opening balance for the new year) should appear on the balance sheet. However, if this is not available, the opening balance can be calculated by adjusting the bank statement balance at the start of the new year by those payments and receipts from the previous year that were unchecked against a statement by the year-end:

- Where the statement balance is in credit, reduce by the unpresented payments and increase by the unpresented receipts.
- Where the statement balance is overdrawn, increase the overdraft by the unpresented payments and reduce by the unpresented receipts.

The correct opening balance cannot be established until all payments and receipts from an earlier period have cleared.

Top Tips

(which may help to resolve the difference if the bank statement and cashbook do not reconcile)

- Find the difference between 'balance at end of month' figures. Is that figure familiar? For example, a regular standing order that had appeared twice in the current month.
- Is the difference divisible by nine? If so, it is likely that somewhere in the cashbook, figures have been transposed.
- Check that the opening balance has been recorded correctly.
- Check for outstanding cheques and receipts from previous months that still have not been presented to the bank.
- If all else fails, recheck the bank statement against the cash analysis entries.

Preparation for the Financial Year-End

Computer programs and spreadsheets remove the need for manual calculations. The cashbook is limited to records of completed financial transactions. Therefore, the summarised figures are not ready for use in the profit and loss account and balance sheet until certain adjustments have been made.

Bank Statements

The cashbook bank balances must be reconciled to the bank statement at the year-end on all accounts.

Recommended Routine

- Deal with any unpresented (over six months old) or duplicated items by making a reverse entry in the cashbook. This may be done by entering the same transaction details in red (or in brackets).
- Reconcile the petty cash box to the petty cashbook.
- Keep photocopies of the closing bank statement bank statements in the statement folder for future reference.

Trade Debtors

List the closing debtors, customers owing money to the business as at the year-end.

Recommended Routine

- Analyse the debtors so that totals may be used to adjust the relevant sales/receipt columns.

Trade Creditors

List the closing creditors, or suppliers owed money by the business, as at the year-end.

Recommended Routine

- Analyse the creditors so that totals may be used to adjust the relevant payments/expenditure columns.
- In addition to trade creditors, there may be PAYE due to HMRC at the year relating to the final months (or quarter) payroll. There may also be costs not yet invoiced that relate to the closing year (e.g., accounting fees).
- See 'Debtors' and 'Current Liabilities' in Chapter 10.

Year-End Cashbook Summaries and Other Adjustments

A summary of the cashbook should be produced including all bank and petty cash accounts and adjusted for opening and closing debtors and creditors.

Recommended Routine

The cashbook example in Diagrams 4.9a and 4.9b shows monthly totals but an annual summary is needed at the year-end. This can either be done by producing a summary of 12 months at the year-end or a separate running total may be done month by month. Again, using a spreadsheet makes light work of either option once the initial set-up has been done.

The annual totals may then be adjusted by deducting the opening debtors/creditors and adding the closing debtors/creditors. However, the closing debtors/creditors must remain identifiable, not simply be added in; they will be required again in a year's time, for subtraction as opening debtors/creditors in next year's account.

The Next Step for Financial Accounts

The next step will depend upon the nature and complexity of the business. For a sole trader operating a very simple business, the next step may be to use the self-assessment online at www.gov.uk, but for a farming business this is only one element of the financial year-end procedure, more details can be found in Chapter 8.

Extending the Use of the Basic Spreadsheet

In this chapter, we have concentrated on routine accounting for VAT and income tax purposes using a basic spreadsheet which replicates the traditional cash analysis book; at this level, the spreadsheet will save time and the presentation of the information should be an improvement on a handwritten ledger. However, the spreadsheet has plenty of scope for development: refinements can be added to automate some of the procedures outlined above. Extra columns may be added for coding, to facilitate more detailed analysis, if required.

As skills develop and new features of spreadsheets are learned, so the cashbook can also be developed. A cash-flow budget (see example in Chapter 11) can be set up in a separate worksheet within the same spreadsheet as the cashbook and 'actual' monthly totals from the cash analysis spreadsheet may be copied into the cash-flow budget. The next step would be to automate this process by referencing cells from the cashbook to the budget.

Spreadsheets are an invaluable tool for anyone involved in recording and analysing data. Agricultural training groups and further education colleges may run short courses in computer skills where there is sufficient demand or an online search will find a range of online courses using Excel.

Sales & Receipts	February 20xx													
		Total	VAT	Analysis	Total unpresented									
Cross Checks		22630.41	22630.41	22630.41	0.00		1	2	3	4	5	6	7	
Set Headings							VAT Recording Headings			Analysis Headings – Enterprise Income				
	TOTALS	21307.92	21307.92			1322.49	21697.87	719.21	213.33	18486.89	0.00	0.00	2000.00	
Date	Detail	Amount	To Bank	REF	Cleared	Contra	Net Amount VAT Outputs	Outside Scope of VAT	Vat charged on sales	Milk	Cull Cows	Calves	Fat Lambs	
12/02/**	Burgett			Contra		120.00	100.00		20.00					
15/02/**	Car & Van Centre			Contra		960.00	800.00		160.00					
03/02/**	P James	100.00		256	*		100.00		0.00					
03/02/**	Machinery Group	120.00	220.00	256	*		100.00		20.00					
13/02/**	The Electricity Company	44.31	44.31	257	*		44.31		0.00					
08/02/**	Faraway Market	1904.00	1904.00	BACS	*	96.00	2000.00		0.00				2000.00	
13/02/**	HMRC VAT	719.21	719.21	BACS	*			719.21	0.00					
15/02/**	Milk Company	18420.40	18420.40	BACS	*	66.49	18486.89		0.00	18486.89				
	Fuel Scale Charge					80.00	66.67		13.33					

8	9	10	11	12	13	14	15	16	17	18	19	20	21
						Other Income							Comment
0.00	0.00	0.00	100.00	0.00	100.00	44.31	0.00	900.00	0.00	0.00	719.21	66.67	
Cull Sheep	Wool	Wheat	Straw	Potatoes	Contracting	Sundry Income	Support Payments & Grants	Capital Sales	Other Income	Private Income	VAT refund from HMRC	Fuel Scale Charge	
								100.00					Chain Saw Trade-in
								800.00					Pick-up Trade-in
			100.00										
					100.00								
						44.31							Wayleave
													26 Lambs
											719.21		
													January Milk
												66.67	

Cashbook Example of Expenditure

Expenditure		February 20xx													
		Total		VAT	Analysis		Total Unpresented								
	Cross Checks	25032.94		25032.94	25032.94		5234.64		1	2	3	4	5	6	7
	Set Headings					VAT Recording Headings			Analysis Headings – Variable Costs						
		Totals		23710.45	1322.49	14222.32	6931.01	3879.61	2613.50	0.00	0.00	427.41	1412.00	0.00	0.00
Date	Detail	REF	Cleared	Amount	Contra	Net Amount VAT Inputs	Outside Scope of VAT	VAT to reclaim	Feed	Livestock Purchase	Vet & Med	Stock Sundries	Seed, Ferts & Sprays	Crop Sundries	Contract, hire & haulage
03/02/**	Feed Company	101	*	1710.00		1710.00		0.00	1710.00						
03/02/**	Digger Company	102	*	1560.00		1300.00		260.00							
03/02/**	Fertilizer Company	103	*	1320.00		1100.00		220.00					1100.00		
03/02/**	Spray & Seed Company	104	*	374.40		312.00		62.40					312.00		
05/02/**	Stationary Store	105	*	10.48		8.73		1.75							
05/02/**	Electricity Co	106	*	668.51		557.09		111.42							
10/02/**	TelCo	107	*	208.10		115.61	69.37	23.12							
10/02/**	Faraway Market	108	*	156.00		156.00		0.00	156.00						
10/02/**	Insurance Company	109	*	1062.62		1062.62		0.00							
12/02/**	Farmers Co op	110		787.72		761.65	23.24	2.83	747.50						
12/02/**	Ford Water	111	*	417.02		417.02		0.00							
12/02/**	Burgett	112	*	360.00	120.00	400.00		80.00							
15/02/**	Fuel Company	113	*	662.09		551.74		110.35							
15/02/**	Credit Card Co Ltd	114		386.92		24.60	357.40	4.92							
15/02/**	Car & Van Centre	115		3980.00	960.00	4000.00	140.00	800.00							
15/02/**	HMRC PAYE	116	*	800.00			800.00	0.00							
15/02/**	James Smith	117		80.00		80.00		0.00							
15/02/**	FG Equipment Finance	118	*	3077.34		1000.00		2077.34							
08/02/**	Faraway Market	CONTRA			96.00	80.00		16.00				80.00			
15/02/**	Milk Company	CONTRA			66.49	55.41		11.08				55.41			
12/02/**	Agricultural Finance Company	DD	*	240.00		200.00		40.00							
20/02/**	NIC	DD	*	10.00			10.00	0.00							
20/02/**	Peter Field	BACS	*	2012.40			2012.40	0.00							
20/02/**	Fred Smith	BACS	*	1913.60			1913.60	0.00							
25/02/**	Bulling Cow	DD	*	350.40		292.00		58.40				292.00			
25/02/**	Bank Charges	DD	*	37.85		37.85		0.00							
28/02/**	Life Insurance Co	SO	*	25.00			25.00	0.00							
28/02/**	J Farmer – Private A/c	SO	*	1500.00			1500.00	0.00							
	Fuel scale charge				80.00		80.00	0.00							

8	9	10	11	12	13	14	15	16	17	18	19	20
Analysis Headings – Overhead Costs							Other Costs					Comment
1202.62	1108.83	118.75	124.34	0.00	4726.00	417.02	237.85	6700.00	1985.01	0.00	80.00	
Mach & Veh Rprs, Veh tax	Fuel, oil & Electricity	Property Rprs & Consumables	Office & Prof Fees	Insurance	Wages & PAYE	Rent, rates, water	Bank Charges, Interest & Finance Chges	Capital	Private	VAT to Repay	Fuel Scale Charge	
												4.1t Dairy Cows £820 DC; 2.2t £440 DYS; 2.25t £450 Sheep
								1300.00				
												10t Fert. @ £110/t Grassland
												Chems Potatoes £110; Chems Grassland £22; Seed W.Wheat £180
			8.73									
	557.09											
			115.61						69.37			2/3rds business/1/3rd private split agreed with HMRC
												Hay for Dairy Young Stock
1062.62												Private car – check fuel scale charge
		14.15							23.24			Wheat feed – Dairy Cow Feed
						417.02						
								400.00				New brush cutter less old trade-in
	551.74											
		24.60							357.40			
140.00								4000.00				Landrover Comm. PF02 HKJ
					800.00							
		80.00										
								1000.00				Bailey Trailer
							200.00					
									10.00			
					2012.40							
					1913.60							
												Dairy Cows
							37.85					
									25.00			
									1500.00			
											80.00	

Chapter 5

From Manual to Computerised Accounts

In the previous chapter, we looked at cash-based accounting systems, which can be done manually on paper, or using a spreadsheet, which is the first step towards using a computerised accounting system. Although figures may be totalled automatically using a spreadsheet and it is possible to produce figures to form a set of accounts, there is still margin for error and the cash-based system has limitations for most businesses. In this chapter, we give an overview of moving from a manual system to a computerised system.

As a business grows and either expands into different enterprises, increases turnover or the need for cash flow and management accounts increases, then it may be the time to consider changing from a manual system to a computerised accounts system that will, if set up logically and with some forethought, provide the business with up-to-date accounts and reports. These reports are often required by consultants and bank managers for management or cash monitoring purposes.

Choosing a Suitable Accounting Software Package

Once the decision to computerise the accounts has been made, then it is necessary to choose a suitable system. There is a wide choice of software packages and they vary in their complexity. Cloud-hosted accounting software is becoming popular. A good broadband connection is essential. For a monthly subscription, data is held off-site, which facilitates shared and mobile access. Generally, bespoke agricultural packages are designed for ease of use; however, where a commercial package is used, a basic knowledge of double-entry bookkeeping is helpful. While it is essential to be able to deal with VAT and produce sufficient financial information for the statutory year-end accounts, the program will also have other useful functions. All systems, correctly set up, should be able to produce listings and reports that will satisfy the accountants; however, it is the ability to produce useful management information that is one of the greatest benefits of using a computerised accounts program.

Commercial accounts programs, which are used for many businesses throughout the UK, can be adapted for a farming business. They are usually fine for routine accounting, such as VAT, but they are limited when it comes to agricultural management accounts. The main limitation is the inability to deal with lengthy production cycles; for example, the production cycle of an arable farm may span two or three financial years. A crop may be planted in year one, grown and harvested in year two but may not be completely paid for until year three.

Choosing a commercial accounts program may seem cheap in comparison to software written specifically for the agricultural market, and will 'do the job', but in order to get full benefit from the accounts, it may be better to pay more and go for software produced by a company that understands the agricultural market and that provides a software support line staffed by people who understand what you are talking about.

Accounts Software Systems

There are several recognised software systems produced for the agricultural and rural estate market that can accommodate the quirks and vagaries of farm and estate accounts. Some specialist software companies offer a 'cut down' or cashbook system that mirrors the operation of the manual cashbook, dealing with routine accounts with the addition of some simple management reports. These systems often provide a useful stepping stone from the manual system up to a full accounts system. See Appendix 1 for software suppliers.

Setting up the System

Once the choice of system has been made, the package usually includes a day of training and setting up the new system. The last set of annual accounts in conjunction with the current cash analysis headings is a good starting point in the design of the new system. Column headings will become codes, usually referred to as nominal codes or headings, generally numeric but sometimes descriptive headings are used. The new system is not limited by length of cash analysis paper. So where the 'Office Expenses' column may have included costs for telephones, stationery, computer costs and sundry office costs, these can all be individually monitored by effectively giving them a column to themselves, all of which can be either reported individually or grouped together to give a total.

Many systems are based on a 'Chart of Accounts', which refers to the overall structure of the accounts in which the nominal codes/headings are grouped. Diagram 5.1 shows a typical 'chart of accounts', this layout is based on gross margin format but other packages may differ.

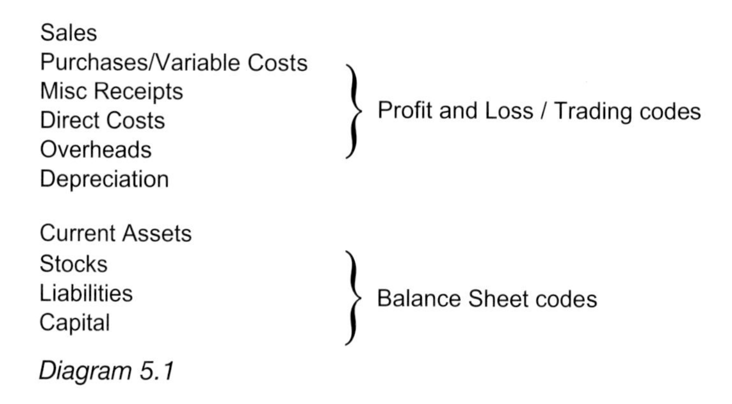

Diagram 5.1

All systems rely upon logical and systematic grouping of codes within each section if accurate reports are to be produced. Within sales there may be subsections for arable income and livestock income, then as many further subsections (codes) as required – under arable: wheat, barley, oilseed rape, etc., and for livestock: milk, calves, lambs, wool, etc. – but keeping within the sales code number range. Similarly, a numeric range should be used to code purchases (variable costs), and another range for overheads. If the business diversifies into a new enterprise, more codes can be added if required in the future.

All systems can be run on either a cashbook or invoice basis, i.e., invoices

CONDENSED LIST OF NOMINAL CODES

Code	Name	Vat Code	Type	Code	Name	Vat Code	Type
SALES							
Arable Income							
00010	Wheat	Z	Normal	00020	Barley	Z	Normal
00030	Grass Seed	Z	Normal	00070	Straw	Z	Normal
00080	Hay	Z	Normal				
Livestock Income							
00100	Cattle	Z	Normal	00110	Lambs	Z	Normal
00120	Wool	S	Normal	00130	Culls	Z	Normal
VALUATION CHANGE							
01799	Valuation Change	O	Valuation				
PURCHASES							
00200	Seeds	Z	Normal	00210	Compound Fertilisers	S	Normal
00220	Nitrogen Fertilisers	S	Normal	00230	Sprays	S	Normal
00250	Marketing Costs	S	Normal	00270	Feed Purchases	Z	Normal
00280	Livestock Purchases	Z	Normal	00290	Vet & Med	S	Normal
00300	Haulage	S	Normal	00310	Contractors costs	S	Normal
00320	Crop Sundries	S	Normal	00330	Stock Sundries	S	Normal
MISC. RECEIPTS							
00600	Misc Revenue	S	Normal	00610	Contract Income	S	Normal
00620	Rental Income	E	Normal	00630	VAT Fuel Scale Charge	S	Normal
00640	SFP	O	Normal				

Diagram 5.2

are entered either when they have been paid or when they are received and then paid off at a later date. The basis that is chosen may depend on the VAT status of the business, ‘cash or invoice’ or how it has always been done in the past. As a business grows and cash flow monitoring becomes more important, then entering invoices as they are received allows the user to see the creditor (money owed by the business) or debtor (to the business) status of the business at any time.

Computerised systems should allow the user to benefit from some or all of the following features, which are discussed in more detail in Chapter 6.

- Entering a multi-lined invoice and allocating it to several different codes.
- Allocating an invoice to different enterprises/departments.
- Entering a contra invoice – for example, purchase of machinery with trade-in.

- Entering an HP Agreement and setting up a liability code (long-term creditor) within the balance sheet.
- Using a journal entry to move values between codes.

Reports

All computerised double-entry systems include the facility to generate reports of debtors and creditors at any date, the report giving the total outstanding balance for each trader together with an aged analysis – up to 30 days, 30–60 days or over 60 days – so that those traders who owe money for which period can be easily identified.

VAT: All systems should allow the user to print off a VAT report, which usually replicates the VAT 100 form, together with detailed listings of the transactions that make up the totals shown on the VAT 100. Detailed reports are essential in the event of a VAT inspection by HMRC. Some systems also have the facility to submit the VAT return online to HMRC.

Activity by code: Reports listing transactions by date range, either on an invoice basis (using the invoice date) or a cash basis, showing what has been paid in the period can be generated. A cash based report is useful for cash flow monitoring, as it shows revenue received and payments made from the bank account and the effect that it would have on the business's borrowings. Most budgeting is done on a cash flow basis for this reason.

Enterprise reports: Where the system allows, the allocation of a transaction to an enterprise/department gives another level of analysis to the accounts. While it is important to know the global picture of the accounts, it is useful to be able to report on the constituent parts of the business. Farms and estates may have different enterprises – for example, arable, cattle, rental – and these may be further split down, so it is possible to report on the wheat, barley and oilseed rape crops that make up the arable enterprise, and the beef and young stock parts of the cattle enterprise. Allocating a cost or receipt to an enterprise is done when the initial purchase or sales invoice is entered. Because the enterprise year for crops can span 18 months and therefore across the financial year-end, most farm accounts software will allow the user to report on enterprise basis, unlike general accounting software, where the management reporting is usually restricted to a 12-month cycle.

Analysis/job code: As a further level of analysis, some systems employ 'analysis/job' codes. These are to assist more accurate costing. Normally these codes are used to allocate costs to machinery or properties and again, the specific code is allocated when the invoice is first entered on the system.

Fixed assets: It is usual for the accounts software to allow the entry of the fixed assets of the business – for example, farm building, tractors, plant and machinery, office equipment – into the accounts so that when machinery is bought and sold it can be added or removed from the Fixed Asset Register. When entering the assets, the user will be asked its purchase price, what depreciation type (straight line or reducing balance) and by what percentage to depreciate the asset. Taking this information, the software should be able to calculate the depreciation value for the asset for the year and post this into the accounts. Reporting on the Fixed Asset Register should show the original value, current values and annual depreciation for each asset; this information is transferred to the profit and loss account at the year-end. See also Chapter 8.

Budgeting: Before the start of the financial year, most businesses will prepare a budget, usually on a cash basis in order to try to forecast the bank borrowing for the next 12 months. These budgets, once entered into the accounts, allow reports on actual transactions against what was budgeted and will report the variance. This is particularly useful to see if the business is performing as expected. Most consultants and bank managers require cash budgeting so that they can monitor the performance of the business on a monthly or quarterly basis. See also Chapter 11.

Other reports: All accounts software will have a set of standard reports that can be used for year-end reporting for management and accounting purposes. Some systems will allow the user to customise their own reports so that they can be set up to report in a specific code or heading sequence.

Saving the Accounts Data in Different Formats

It is usually possible to save a report generated in a software package in different formats. For example, a report of nominal activity for all the codes relating to crop inputs may be generated for a growing season and exported to an excel spreadsheet for further analysis or it could be saved as a .pdf (portable document format) to enable the report to be attached to an email.

Often the idea of changing from a manual to a computerised system fills the bookkeeper with horror, but with proper training on a suitable system, there are long-term benefits for all concerned.

Top Tips

- Check the computer's specification and operating system is compatible with running the software.
- Leave gaps in a numeric sequence for possible additional codes in the future.
- Use the previous year's accounts as a guide to account headings/ codes.
- Get professional help in setting up the system.
- Explore the management reporting tools as this can save time – play with it!

Chapter 6

Double-Entry Computerised Accounts

Earlier chapters have concentrated on simple cashbook systems and have discussed making the changeover from a manual to a computerised system. Single-entry systems, where the entries are made once the invoice has been received and the payment has been made, are adequate for many small businesses, where there are only a small number of sales and purchases in a year and it is easy to keep track of outstanding invoices. However, in larger businesses where there are more transactions, a cashbook system may not give the adequate financial and management control, which may mean that best use is not being made of resources. A double-entry system allows the bookkeeper to keep a track of debtors, money owed to the business, and creditors, money owed by the business, monitor cash flow and produce traditional gross margin reports for individual enterprises.

Double-entry bookkeeping involves making two entries in the accounts for each transaction.

- The first entry is the purchase or sales invoice entered onto the supplier or customer ledger, automatically updating the nominal ledger with an additional creditor or a debtor.
- The second is the payment from or receipt into the bank account in respect of that invoice, automatically updating the nominal ledger with the reduction in the total outstanding due to creditors or due from debtors.

In a cash accounting (single-entry) system, one record is made combining the payment or receipt and the analysis of the transaction.

Setting up a Computerised Double-Entry System

In times past, double-entry bookkeeping involved making two entries into separate handwritten ledgers. With the advent of computerised accounts the concept is easier to understand and put into practice. When an invoice is entered on the system via the supplier/purchase ledger, or the customer/sales ledger, the computer automatically updates the nominal ledger, where the double-entry accounting record is maintained.

The double-entry system comprises three ledgers:

- **Supplier or purchase ledger:** Details of suppliers and purchase invoices are entered here.
- **Customer or sales ledger:** Details of customers and sales invoices are entered in this ledger.
- **Nominal ledger:** This is the record of the double-entry accounts.

Separate records for all traders, (*suppliers and customers)*, are set up, as well as the nominal account code structure as described in Chapter 5. A record for each asset is set up as per the closing net book value from the last set of accounts. Reconciled bank balances are required; however, a complete ‘opening trial balance’ (closing trial balance from the last set of accounts) is not essential at the set-up stage. This may be obtained from the accountant and entered later, a task that may require guidance from the accountant or software supplier.

In addition to accounting functions, it is possible to set up additional analysis functions for management reports, such as enterprise gross margins and cash-flow control.

Data Entry

Different computerised accounts systems will have differing data entry screens. The main types of data entry are:

- Entering invoices onto the supplier or customer ledger.
- Entering payments or receipts into the bank account/s.
- Entering accounting adjustments using a journal into the nominal ledger.

Regarding *purchase/supplier invoice*, details of the invoice are entered on the supplier's account in the supplier ledger, analysing the VAT and the net value of the input to the appropriate nominal code/s. This invoice will remain outstanding on the supplier ledger until the payment is entered on the system.

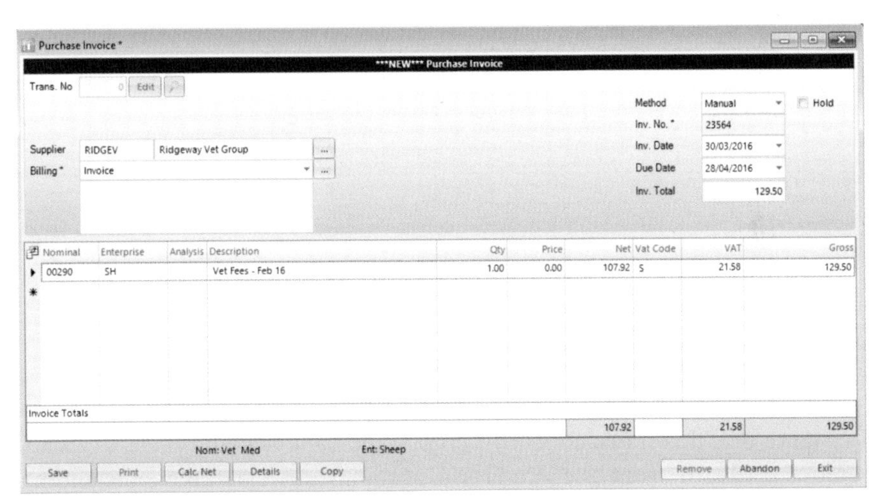

Figure 6.1 Purchase invoice

In Figure 6.1, this entry will post £107.92 to the code 290 – Vet & Med, £21.58 to the VAT inputs account and £129.50 to the supplier ledger under Ridgeway Vet Group. If management requires that, for example, cattle vet and med be separated from pig vet and med, it is possible to use two lines, with appropriate analysis codes in the right-hand column: it is immaterial where the claimable VAT is shown. This will also show when payment is due, allowing the business to take advantage of the credit terms allowed by the supplier.

Purchase/supplier payments are not made in the supplier ledger but in a separate bank data entry point, where the bank account may be selected (i.e., current, petty cash, etc.) as well as the method of payment (cheque, BACS or direct debit), the date the payment is made and the reference/ cheque number. Most systems will allow 'part payment' of an invoice or to pay an amount 'on account'.

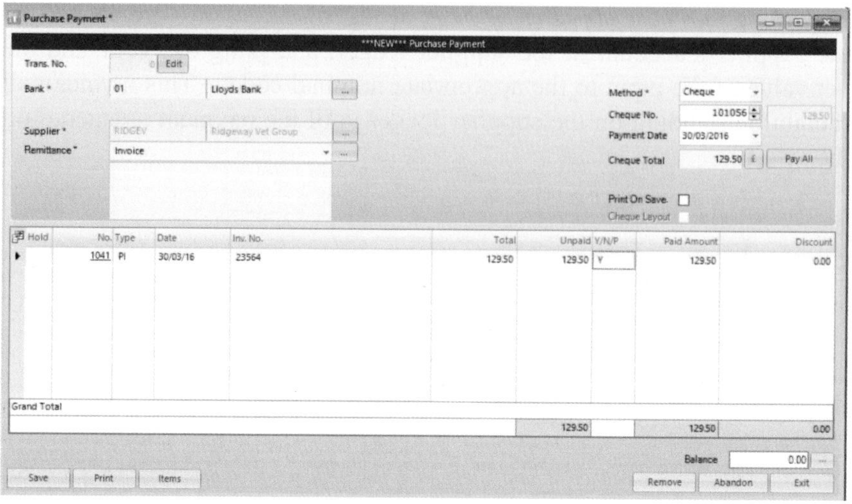

Purchase Payment *

NEW Purchase Payment

Trans. No. 0 Edit
Bank * 01 Lloyds Bank
Supplier * RIDGEV Ridgeway Vet Group
Remittance * Invoice

Method * Cheque
Cheque No. 101056
Payment Date 30/03/2016
Cheque Total 129.50 £ Pay All
Print On Save
Cheque Layout

Hold	No.	Type	Date	Inv. No.	Total	Unpaid	Y/N/P	Paid Amount	Discount
	1041	PI	30/03/16	23564	129.50	129.50	Y	129.50	0.00
Grand Total						129.50		129.50	0.00

Balance 0.00

Save Print Items Remove Abandon Exit

Figure 6.2 Purchase payment by bank account

In Figure 6.2, this transaction will charge the bank account code £129.50, remove £129.50 from the creditor's account and clear the outstanding Ridgeway Vet Group account as the supplier ledger report in Figure 6.3 shows.

Sales/customer invoice details are entered in the customer account on the customer ledger, analysing the VAT and the net value of the output to

Supplier Ledger

From 01/01/2016 to 30/04/2016

RIDGEV Ridgeway Vet Group 12 Mth. Activity 107.92 Opn. Bal. 0.00

Hold	Date	Ty	No.	Reference	P.Invoice	Payments	Discount	Balance	Payment No.	Statement No.
	30/03/16	PI	1041	23564	129.50	0.00	0.00	129.50	1043	-1
	30/03/16	PP	1043	101056	0.00	129.50	0.00	0.00	1043	
				Total	**129.50**	**129.50**	**0.00**	**0.00**		
				Grand Total	**129.50**	**129.50**	**0.00**			

Figure 6.3 Supplier ledger

the appropriate nominal code/s. This invoice will remain outstanding on the customer ledger until the receipt is entered on the system.

In Figure 6.4, this entry will post £2,000 to code 610 – Contracting Income, £400 to the VAT outputs account and £2,400 to the customer

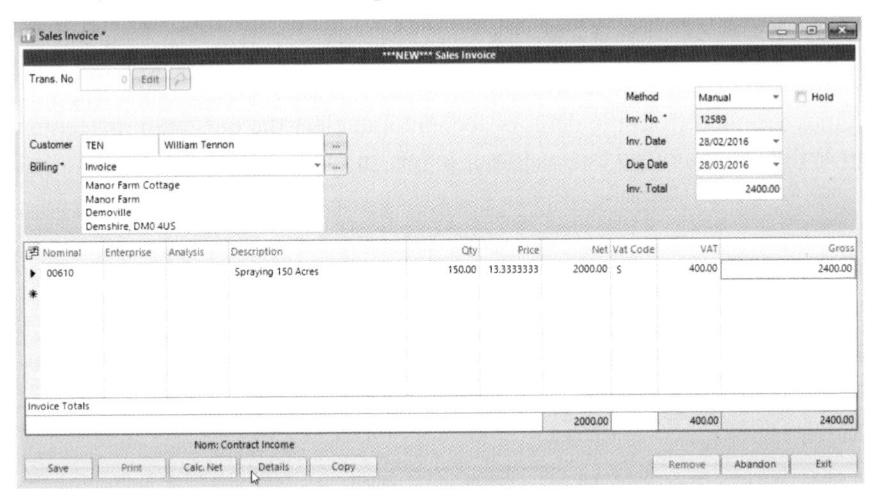

Figure 6.4 Sales invoice

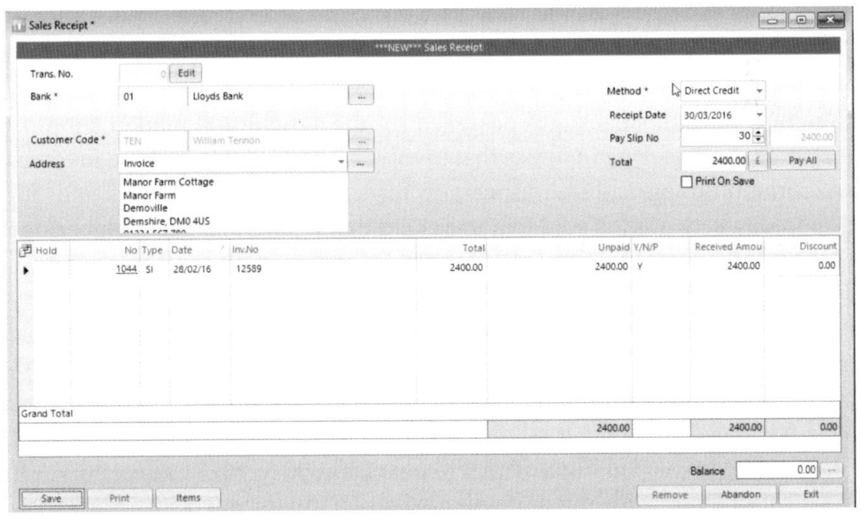

Figure 6.5 Sales receipt by bank account

ledger under William Tennon. The due date for payment by William Tennon is also recorded.

Sales/customer receipts are not made in the customer ledger but in a separate bank data entry point, as described in purchase payments above. When the payment is received, either by cheque or direct to the bank account, the following transaction would be entered.

In Figure 6.5, this transaction will add £2,400 to the bank account, remove £2,400 from the debtors account and clear the outstanding account of William Tennon as the customer ledger in Figure 6.6 shows.

Customer Ledger
From 01/01/2016 to 30/04/2016

TEN	William Tennon			12 Mth. Activity	2000.00	Opn. Bal.	0.00	Payment	Statement
Hold	Date Ty	No.	Reference	S.Invoice	Receipts	Discount	Balance	No.	No.
	28/02/16 SI	1044	12589	2400.00	0.00	0.00	2400.00	1046	-1
	30/03/16 SR	1046	30	0.00	2400.00	0.00	0.00	1046	
			Total	2400.00	2400.00	0.00	0.00		
			Grand Total	2400.00	2400.00	0.00			

Figure 6.6 Customer ledger

Contra Entries

Computer accounts packages designed for the agricultural market simplify the process of entering invoices that involve a 'contra' entry. These invoices are common in agriculture, especially for trade-ins, auction sales and, for example, where corn is sold to a merchant that in turn charges weighbridge charges, HGCA Levy, etc., all of which is included on the same invoice. These are termed 'self-billing' invoices, where the customer is actually invoicing your business on your behalf. The milk cheque is a prime example of this. The self-billing invoice from the dairy advises the business as supplier of the number of litres of milk supplied in the month, what deductions have been made, and the total that is being credited to the bank.

The sales invoice in Figure 6.7 will post £10,000 to wheat sales, no VAT, as wheat sales are zero-rated; it will also post £150 to code 250 – Marketing Costs, being the HGCA Levy and the weighbridge charges, charged to the

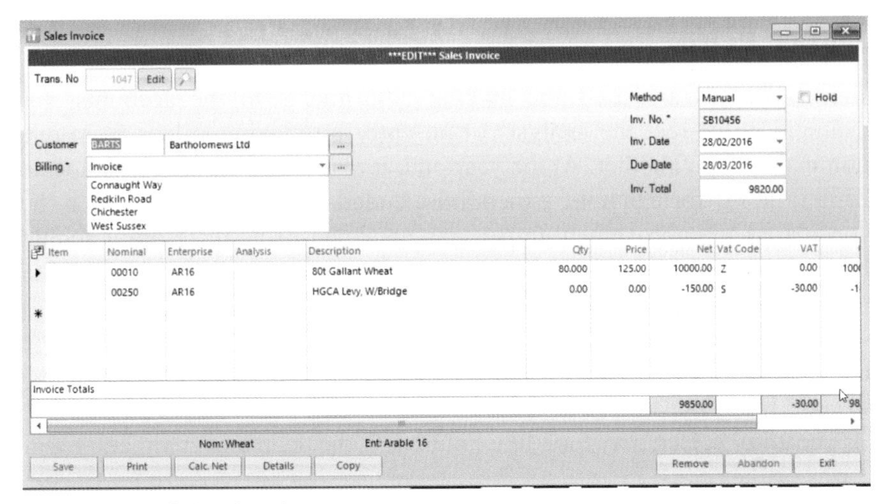

Figure 6.7 Sales invoice

CUSTOMER LEDGER - DETAILED

From 01/01/2016 to 30/04/2016

Trader: BARTS		Bartholomews Ltd		12 Mth. Activity	9850.00		Opening Balance		0.00
Type No.		Date Hold Bank / Chq No.	Inv. Reference			Discount	Total	Ac Bal	Pay No.
Nom	Ent	Analysis	Description	Qty	Amount	VAT		Ref	
SI 1047		28/02/16	SB10456						
00010	AR16		80t Gallant Wheat	80.00	10000.00	0.00			
00250	AR16		HGCA Levy, W/Bridge		-150.00	-30.00			
					9850.00	-30.00	9820.00	**9820.00**	0
							Closing Balance		**9820.00**

Figure 6.8 Customer ledger – detailed

business by Bartholomews and £30 to the VAT input code. The balance of £9,820 will be posted to the customer ledger under Bartholomews as shown in Figure 6.8.

In the examples above, only one invoice has been paid off with each payment but, as with a single-entry system where one cheque number would be assigned to several invoices, it is possible to pay off several invoices at one time and most systems will have this facility. The accounting is exactly the same, i.e., the creditor's outstanding balance is being reduced by the value of the cheque.

Journal Entries

Journal entries are used for making adjustments to the accounts; for example, to correct the analysis of an entry by moving values from one nominal code to another. At the year-end, it may be necessary to reallocate certain items; for example, expenditure coded to property repairs may be considered an improvement rather than an expense and therefore it should be capitalised (it then becomes a balance sheet item). A journal is used to transfer the sum from a purchase code to a fixed asset code.

The following example of a journal entry (Figure 6.9) shows the reallocation of £1,050 of expenditure from code 1110 (Repairs – Vehicles) to code 1100 (Repairs – Tractors).

The entry screen may look like Figure 6.9, the debit and credit entry will always balance.

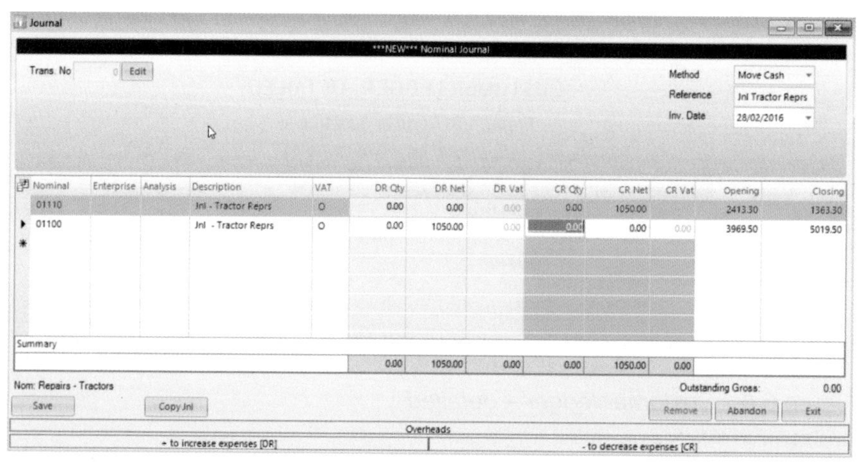

Figure 6.9 Journal entry

Accruals and Prepayments

These are specific types of journal entries that are used mainly at year-end to either bring into the accounts expenses/receipts that have occurred after the year-end, but which refer to the present accounting year, or, conversely, transactions that have taken place that refer to the new accounting period.

A typical example of a prepayment is that of an annual subscription for £72 that has been paid for the 12 months starting 1 October in a set of accounts with a year-end on 31 March. Clearly, only half of the subscription (six months/October to March) refers to this year's accounts, so a journal is needed to move 6/12th (£36) of the value to next year.

The journal should be dated at the last day of the financial year, for example 31 March and will move £36 from 'Subscriptions' to a balance sheet code called 'Prepayments'. A reversing journal dated 1 April (the first day of the new year) will then move £36 from 'Prepayments' back to 'Subscriptions' (Diagram 6.1).

Year-End Prepayment Journal – 31st March

Code	Code	Debit	Credit
Subscriptions	1470	£36	
Prepayments	7000		£36

Year-Start Prepayment Journal (Reversing) – 1st April

Code	Code	Debit	Credit
Subscriptions	1470		£36
Prepayments	7000	£36	

Diagram 6.1

An accrual journal will bring the cost of an invoice dated after the year-end back into the current year's accounts. The example shown in Diagram 6.2

Year-End Accrual Journal – 31st March

Code	Code	Debit	Credit
Repairs-Tractors	1100	£1000	
Accruals	7100		£1000

Year-End Accrual Journal (Reversing) – 1st April

Code	Code	Debit	Credit
Repairs-Tractors	1100		£1000
Accruals	7100	£1000	

Diagram 6.2

is an invoice for tractor repairs for £1,000 which is dated 30 April but refers to work actually done in March.

The major farm accounts packages will allow the user to do both the year-end journal and the reversing year-start journal in one entry, thus reducing the user's workload.

Bank Reconciliations

The method is the same as shown in Chapter 4, although the computer does most of the work. The statement reference, closing date and bank balance for the current period are entered manually. The balance carried forward – shown on the reconciliation screen – should be checked against the opening bank statement balance for the current period. The computer then presents a screen of recorded transactions on which corresponding entries can be identified on the bank statement and 'ticked off'. Any missing items from the accounts such as bank charges may be entered and ticked off. Uncleared cheques will remain live (on the screen) until the next time the account is reconciled, at which point they will appear on the bank statement.

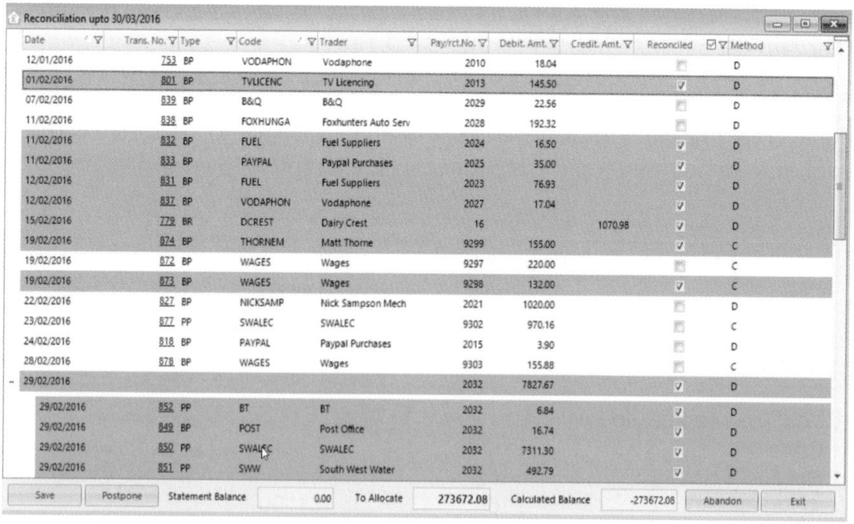

Reconciliation upto 30/03/2016

Date	Trans. No.	Type	Code	Trader	Pay/rct.No.	Debit. Amt.	Credit. Amt.	Reconciled	Method
12/01/2016	753	BP	VODAPHON	Vodaphone	2010	18.04			D
01/02/2016	801	BP	TVLICENC	TV Licencing	2013	145.50		✓	D
07/02/2016	839	BP	B&Q	B&Q	2029	22.56			D
11/02/2016	838	BP	FOXHUNGA	Foxhunters Auto Serv	2028	192.32			D
11/02/2016	832	BP	FUEL	Fuel Suppliers	2024	16.50		✓	D
11/02/2016	833	BP	PAYPAL	Paypal Purchases	2025	35.00		✓	D
12/02/2016	831	BP	FUEL	Fuel Suppliers	2023	76.93		✓	D
12/02/2016	837	BP	VODAPHON	Vodaphone	2027	17.04		✓	D
15/02/2016	779	BR	DCREST	Dairy Crest	16		1070.98	✓	D
19/02/2016	874	BP	THORNEM	Matt Thorne	9299	155.00		✓	C
19/02/2016	872	BP	WAGES	Wages	9297	220.00			C
19/02/2016	873	BP	WAGES	Wages	9298	132.00		✓	C
22/02/2016	827	BP	NICKSAMP	Nick Sampson Mech	2021	1020.00			D
23/02/2016	877	PP	SWALEC	SWALEC	9302	970.16			C
24/02/2016	818	BP	PAYPAL	Paypal Purchases	2015	3.90			D
28/02/2016	878	BP	WAGES	Wages	9303	155.88			C
– 29/02/2016					2032	7827.67		✓	D
29/02/2016	852	PP	BT	BT	2032	6.84		✓	D
29/02/2016	849	BP	POST	Post Office	2032	16.74		✓	D
29/02/2016	850	PP	SWALEC	SWALEC	2032	7311.30		✓	D
29/02/2016	851	PP	SWW	South West Water	2032	492.79		✓	D

Save | Postpone | Statement Balance 0.00 | To Allocate 273672.08 | Calculated Balance -273672.08 | Abandon | Exit

Figure 6.10 Bank reconciliation

Management Accounts

Different levels of accounting are possible using a computerised system, which is a sophisticated database. So far we have discussed the accounting functions of the system but beyond this, software designed for agricultural businesses has the facility to produce reports by enterprise in traditional gross margin format. However, this feature is not automatic; there will be a standard set of reports but, to make the best use of the system, a suitable range of enterprise income codes/headings and related expenditure codes/headings will need to be set up. This is where a day or two of on-site training and set-up is invaluable. The software designers are best placed to advise how their product will serve the reporting needs of the business.

Budgeting

All businesses have to think ahead and the more efficient will commit their plans to an annual budget in order to gauge bank borrowing for the next 12 months. A budget (plan) can be entered into the accounting system before the start of the year. As (actual) data is entered on the system, the true picture can be compared against the budget. A standard 'variance' report can be produced and is commonly used by consultants and bank managers where external monitoring of cash flow is required.

The basics of management reporting and budgeting requirements using spreadsheets are covered in Chapter 11, but with specialist software many of these reports can be programmed to give the same information. However, it is common for stand-alone budgets to be created and monitored using a spreadsheet; the data export facility included in accounting software may save time when updating actual figures against a budget (plan).

Conclusion

While it may seem that using a double-entry system means more work by having to enter an invoice to then pay it off, it actually allows the user to keep an eye on the bank balance, money owed to the business and by the business, without having to do complicated manual or spreadsheet calculations, to prepare the accounts for the year-end in a format that the accountants require and to produce management figures straight from the accounts programme.

Top Tips

- Choose tried and tested accounting software with a reliable helpline that understands the nuances of agricultural accounts.
- Use template or memo functions for repeat transactions.
- Use recurring entry or standing order functions for repeat banking transactions.
- Have checking procedures in place to pick up misanalysis and other mistakes.
- Check data and take backups regularly.

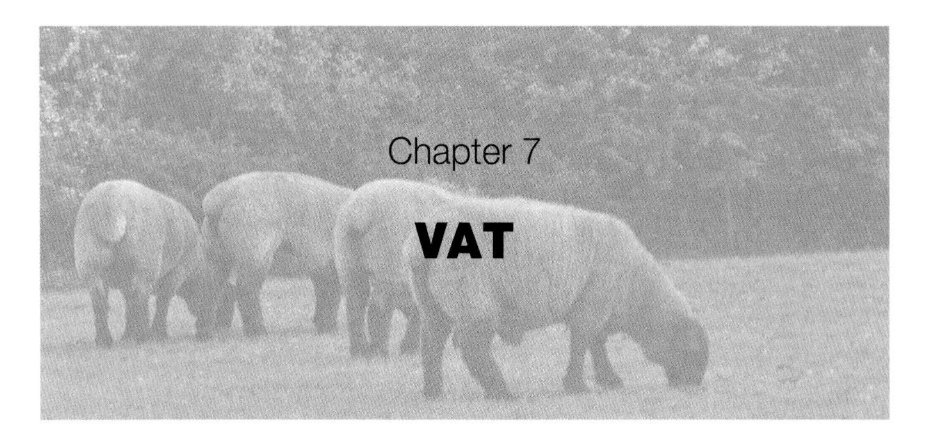

Chapter 7

VAT

Value Added Tax (VAT) is an indirect tax that is imposed on goods and services at each stage of production. Similar taxes are also in place in other countries, both within and outside the European Union (EU). In this chapter, we aim to give an overview of how VAT rules and regulations may affect farming and other small businesses in general. It is a highly complex area and more comprehensive information can be found in the VAT Guide at www.gov.uk.

Transactions within the UK are either subject to standard rated VAT, reduced rate VAT, zero-rated VAT or they may be exempt from VAT or outside the scope of VAT as defined in the VAT Guide. Some of the most common transactions that farming businesses encounter fall into the following groups.

Standard Rate

- Purchase of fertiliser and sprays
- Purchase of straw for bedding
- Contracting
- Fuel.

Reduced Rate

- Purchase of domestic heating fuel
- Sale of wood for use as heating fuel.

Zero Rate

- Sale of crops or animals (as food)
- Purchase of seeds (for food production)
- Purchase of straw for feeding to livestock or horses
- Books and newspapers
- Most public transport
- Water and sewerage services.

Exempt

- Rental income, other than holiday lets
- Education, vocational training
- Insurance
- Stabling
- Rights over land
- Sales of freeholds and leaseholds of land and property (subject to certain exceptions).

Outside the Scope

- Basic Payment Scheme (BPS), other environmental payments and grants
- Wages and salaries
- Bank interest
- Voluntary donation to charities
- Transactions of a private nature
- Transfers of capital, loans, etc.
- Purchases from a business not registered for VAT.

Registration for New Businesses

A new business must keep track of its taxable turnover (taxable supplies for VAT purposes) and at the end of any month, if the value of your taxable supplies in the previous 12 months or less is over the current registration threshold; or at any time, it is expected that the value of taxable supplies in the next 30-day period alone will go over the registration threshold, the business must register for VAT. This must be done within the next 30 days. There are penalties for late registration. An exemption from registration may be applied if it can be shown that the business will not breach the registration limits within the following 12 months.

It is not possible to register for VAT if either of these is true:

- The only goods or services sold are exempt from VAT.
- It is not a business according to the definition that HMRC uses for VAT purposes.

VAT-registered businesses must charge VAT on standard and reduced rate supplies, which is then due to be paid over to HMRC. In return, they can generally reclaim the VAT they have paid on their expenses. They will prepare a VAT return on either a monthly or quarterly basis. If the VAT on taxable supplies (Output VAT) is greater than the VAT on taxable purchases (Input VAT), then the net amount must be paid to HMRC. Alternatively if the Input VAT is higher than the Output VAT, then the business will be due a VAT refund. Most farming businesses are regularly due a VAT refund because a high proportion of their sales are zero-rated.

Example

The cashbook in Diagram 7.1 shows output VAT of £200 and input VAT of £942 for September 2017. A VAT refund of £742 would therefore be due for the month.

Month 1				Expenses		VAT Recording Columns			Analysis Columns			
		Totals		7452.00	0.00	6510.00	0.00	942.00	1800.00	3200.00	60.00	1450.00
Date	Detail	REF	Cleared	Amount	Contra	Net Amount VAT Inputs	Outside Scope of VAT	VAT to reclaim	Seed	Sprays	Fuel	Repairs
03/02/**	Atlas	PO901	*	2640.00		2200.00		440.00		2200.00		
03/02/**	Atlas	PO902	*	1800.00		1800.00		0.00	1800.00			
03/02/**	Collings Bros	PO903	*	1200.00		1000.00		200.00				1000.00
03/02/**	M Weatherhead	PO904	*	540.00		450.00		90.00				450.00
05/02/**	ABC Garages	PO905	*	72.00		60.00		12.00			60.00	
05/02/**	Atlas	PO906	*	1200.00		1000.00		200.00		1000.00		

Diagram 7.1

Month 1				Income		VAT Recording Columns			Analysis Columns			
		Totals		12796.99	0.00	9652.00	2944.99	200.00	8652.00	0.00	2944.99	1000.00
Date	Detail	REF	Cleared	Amount	Contra	Net Amount VAT Outputs	Outside Scope of VAT	VAT to charged on sales	Wheat	Beans	VAT Refund from HMRC	Misc
01/02/**	Wellgrain 30.5T	RO901	*	4575.00		4575.00			4575.00			
03/02/**	Wellgrain 30.2T	RO902	*	4077.00		4077.00			4077.00			
03/02/**	VAT Refund -Dec	RO903	*	2930.00			2930.00				2930.00	
04/02/**	Deposit Interest	RO904	*	14.99			14.99				14.99	
05/02/**	Greens Truck AB02 TYP	RO905	*	1200.00		1000.00		200.00				1000.00

Diagram 7.1 Continued

Partial Exemption

If the business has any form of exempt income, partial exemption should be a constant consideration. The basic principle is that 'Input tax cannot be reclaimed on goods or services that are used in making exempt supplies'. Where a business has only exempt sources of income then no VAT may be recovered from any expenses.

In most cases, it is possible to attribute input tax to either taxable supplies (e.g., fertiliser and spray costs can be attributed to zero-rated crop sales) or exempt supplies (e.g., legal fees for the sale of land). However, there are other expenses, such as office expenses and accountancy fees, that are attributable to both. The VAT on these costs is known as 'residual input tax' and this needs to be split in an equitable method, so that the proportion of residual input tax that is deemed to relate to taxable activities can be recovered, whereas the proportion relating to exempt supplies cannot.

HMRC allows a variety of methods to calculate this split; however the 'standard' method is the most common. The following example (Diagrams 7.2 and 7.3) shows a business with both taxable and exempt supplies:

Sales	£
Taxable supplies Exempt supplies	10,000 8,000
	18,000
Input tax	
Directly attributable to taxable supplies Directly attributable to exempt supplies Not directly attributable to either	1,500 500 300
	2,300

Diagram 7.2

Recoverable proportion of residual input tax is calculated as follows.

$$\frac{\text{Taxable turnover}}{\text{Taxable + exempt turnover}} \quad \text{x} \quad \text{Residual input tax}$$

$$\frac{10{,}000}{18{,}000} \text{ x } 300 = £167$$

Therefore of the total £2,300 input VAT incurred in the month,

1,500 + 167 = 1,667 can be reclaimed and

500 + 133 = 633 cannot

Diagram 7.3

Option to Tax Non-Domestic Properties

One way to overcome the problem of exempt supplies restricting the repayment of input VAT is to consider opting to charge VAT on your properties. This is possible for most non-domestic properties, but advice should always be sought. The basic principle is that you opt to charge VAT on all income from the property, such as rental income, lease premiums or sale proceeds. This then becomes a standard rated supply and all VAT on expenditure relating to that property is therefore recoverable. Once made, the election is irrevocable for 20 years, although there is a 'cooling off period' of three months from the date of the election provided there have been no tax consequences. The option may be exercised separately in relation to each building or discrete area of agricultural land. Therefore if you are only slightly over the *de minimis* limit (see below) and have a VAT registered tenant in one property, you may decide to opt to tax that building to ensure that all VAT relating to your properties can be recovered. The terms of this option are subject to revision and up-to-date advice should be sought from an accountant or HMRC

De Minimis Rules

If the total of all exempt input VAT and exempt residual VAT is less than £625 per month on average *and* 50% of the total input VAT incurred, then *all* input tax can be recovered. In the example above, the exempt VAT (£500) and residual exempt VAT (£133) is more than £625, therefore you must disallow the whole of this £633. These limits are subject to revision, which should be checked at www.gov.uk.

Other special methods may give a more favourable result, therefore consider the method used carefully. For instance, country estates often find that an apportionment based on input tax rather than sales is generally more beneficial due to the high level of exempt income they receive from rented property.

The partial exemption calculation must be carried out for each month or quarter that you prepare a VAT return and in addition there is also an annual calculation. Where VAT has been disallowed during the year and the annual calculation shows that the annual *de minimis* limit has not been reached, then at this point the VAT previously disallowed may be claimed. This adjustment may also result in VAT having to be repaid to HMRC!

Using the Previous Year's Percentage

It is possible to use the final apportionment that was calculated for the previous year as the basis for calculating residual expenditure in the current year. This saves having to carry out the partial exemption calculation each month, but the annual calculation must still be done and any adjustment necessary made at that point.

Details of the information that must be shown on purchase and sales invoices can be found in Chapter 3 and also at www.gov.uk.

VAT Inclusive Invoices

Occasionally VAT inclusive invoices are received; examples include vending machines, other machine services such as photocopiers, and some retail sales. In these cases, in order to claim back the VAT on the purchase, it is necessary to calculate the VAT proportion of the invoice. To work out the VAT exclusive (net) amount, where the VAT rate is 20 per cent, simply divide the gross amount by 1.2.

For example, if the price including VAT is £120, then the price excluding VAT will be:

120 ÷ 1.2 = £100
The VAT element is therefore £120 – £100 = £20

An alternative method is to use a 'VAT fraction' to find the amount of VAT that has been included in a VAT-inclusive total. Multiply the total by the appropriate VAT fraction to arrive at the VAT. To work out the VAT fraction, start with the VAT rate divided by 100, plus the VAT rate and reduce this down by dividing the top and bottom figures by the highest figure possible: 20/120 is therefore reduced to 1/6.

Rate of VAT	VAT fraction
Standard rate – 20%	1/6
Reduced rate – 5%	1/21

Diagram 7.4

Using the VAT fraction of 1/6 VAT included in a total of £120 is as follows:
120 × 1/6 = £20

The Annual Accounting Scheme – Small Businesses General

This scheme is designed to simplify VAT recording for small businesses. Nine monthly payments are made by direct debit – based on an estimate of the amount of VAT due. A tenth payment, two months after the end of the accounting year, is made to balance the account. An application to join the scheme is required, more rules and details are available at www.gov.uk.

The Cash Accounting Scheme – Small Businesses General

This scheme is widely used by small businesses. VAT is reclaimed on a cash basis – i.e., when the payment is made or received rather than on the date of the invoice. It is more beneficial to a business supplying goods and services at the standard rate and paying HMRC, than for a farming business where returns usually result in a refund. Criteria for joining the scheme may be found at www.gov.uk.

The Flat Rate Scheme – Small Businesses General

This is the remaining scheme for small businesses. It is most useful for businesses offering goods and services where standard rate VAT is charged and where VAT reclaimed is relatively low. VAT invoices for sales are produced in the usual way, but no VAT is claimed on purchases and expenses except for certain purchases of capital assets with a VAT inclusive value of £2,000 or more. VAT liability is calculated by applying a flat-rate percentage to the VAT inclusive turnover. The flat-rate percentage depends on the trade sector into which a business falls for the purposes of the scheme. There is a wide spread of applicable percentages ranging from 5% to 14.5%. A calculator to estimate if the scheme may be beneficial and an application form, rules and more details are available at www.gov.uk.

The Agricultural Flat-Rate Scheme

The flat-rate scheme is an alternative to VAT registration for farmers and growers. Instead of being registered for VAT, a producer can apply to the local VAT office for a certificate as a flat-rate farmer. Such producers do not have to submit returns and consequently do not have to account for or reclaim tax. Instead they charge a flat-rate addition (4%) when they sell to VAT-registered customers goods or services that qualify, including VAT zero-rated supplies. This addition is not VAT but acts as compensation for not being able to recover input tax on purchases. It is reclaimable by a VAT-registered customer. An application to join the scheme is required and rules and details are available at www.gov.uk.

Cars and Motoring Expenses

When a car is purchased, generally VAT cannot be reclaimed. There are some exceptions to this; for example, when the car is used mainly as a taxi, for driving instruction or for self-drive hire.

If a car is leased for business purposes, 50% of the VAT paid can normally be reclaimed but it is 100% if the car is used as one of the above or where you can show that it is used exclusively for a business purpose. This is not easy to prove. It may be necessary to show that insurance of the vehicle is restricted to business use plus any other documentary evidence to demonstrate how it meets this requirement.

An anomaly of the VAT system is that it is possible to reclaim all the VAT charged on vehicle repairs and maintenance so long as the business pays for the work and there is some business use of the vehicle, no matter how small. It is generally also possible to recover VAT on all other business motoring expenses such as fleet management or off-street parking, with a VAT invoice to support the claim.

There are special rules for reclaiming VAT on road fuel used for business purposes. If the business pays for road fuel, there are four methods of accounting for VAT.

1. Reclaim all of the VAT – the fuel must only be used for business purposes.
2. Reclaim all of the VAT and pay the appropriate VAT fuel scale charge.
3. Reclaim only the VAT that relates to fuel used for business mileage. Detailed records of your business and private mileage must be kept.
4. Not reclaim any VAT. This can be a useful option where business mileage is low and there is an element of private motoring. If this option is chosen, it must be applied to all vehicles including commercial vehicles.

VAT Fuel Scale Charge

This is a simplified means of taxing the private use of business fuel. If this method is selected, VAT on private fuel is accounted for by reference to VAT Fuel Scale Charge as set in the Budget, the table of rates is available on the HMRC website. Fuel Scale Charge is payable per VAT return period, it is determined by the CO_2 emissions of a vehicle, this can be found on the registration certificate for the vehicle or by clicking on the vehicle enquiry service at www.taxdisc.direct.gov.uk.

VAT Treatment of Shared Machinery

In recent years, it has become more common for farmers to share the purchase of farm machinery. This helps to reduce the cost to each participator as well making economies of scale. However, complications can arise when accounting for the VAT. As far as the supplier of the asset is concerned, there is only one customer, which will either be one of the farmers or to the whole group (if separately registered, e.g., as a syndicate). The VAT invoice will therefore be addressed to that customer and they will be able to reclaim the input tax, subject to the normal rules. Payments made by other members to that person, whether for the initial purchase of the article or its maintenance, are consideration for the right to use the asset. This is a supply of services, and output tax must be accounted for by the registered member or group as appropriate. If the article is eventually resold, output tax is due from whoever originally received the article. Payments made by the registered member or group to the other members in order to pass on the proceeds of the sale are outside the scope of VAT as a profit share.

Repairs and Improvements to Farmhouses

The VAT treatment of land and property transactions is complicated and it pays to seek professional advice before embarking on major renovation projects.

For general repairs and improvements to the farmhouse, HMRC provides specific advice. There are different rules applicable for farmhouses occupied by a sole trader or a partner and those occupied by directors. It is unlikely that any VAT will be recoverable where the occupier is a director. For partners and sole traders, however, the situation is very different. HMRC agreed to the following principles.

First, where VAT is incurred on repairs, maintenance and renovations, 70% of that VAT may be recovered as input tax provided the farm is a normal working farm and the VAT-registered person is actively engaged full-time in running it. Where farming is not a full-time business for the VAT-registered person, input tax claimable is likely to be between 10% and 30% on the grounds that the dominant purpose is a personal one.

Second, where building work is more associated with an alteration (e.g., building an extension) the amount that may be recovered will depend on the purpose of the construction. If the dominant purpose is a business one, then 70% may be claimed. If the dominant purpose is a personal one, HMRC would expect the claim to be 40% or less, and in some cases, depending on the facts, none of the VAT incurred would be recoverable.

Other Farm Buildings

Assuming the farm buildings are non-residential and used for business purposes, as a general rule, 100% of the VAT is recoverable. Care should be taken if buildings are let and a claim to opt to tax them has not been made (see partial exemption above).

Additional advice should be sought if a non-residential building is being converted into residential, there is a change of use or it is a listed building. There are a range of complicated regulations that determine the appropriate rate of VAT that should be charged and it can be difficult to unwind any mistakes. The cost of obtaining advice in advance is likely to be minor compared to the potential cost of any error.

Retention of VAT Records

It is necessary for all records and accounts that provide the basis for VAT returns, to be kept for a maximum of six years so that they are available for inspection by HMRC if required. Digital records are acceptable in pdf or picture format provided all of the relevant information is readable.

Online VAT

From April 2012, all VAT-registered businesses have to submit VAT returns online and pay electronically.

Registration and Activating the Service

Registration with Government Gateway (www.gateway.gov.uk) and selection of the service for VAT Online is required. Only one registration is required, it may be simply a matter of logging in and selecting VAT Online if the business is already registered. It is possible for self-employed administrators to register as agents in order to deal with their clients' VAT returns etc.

VAT Returns

VAT returns should be made by the 7th of the month following the end of the month after the VAT period. Penalties are imposed on businesses that submit late returns.

The current return requires completion of nine boxes:

Box 1 VAT due in the period on sales and other outputs.
Box 2 VAT due in the period on acquisitions from other EU Member States.
Box 3 Total VAT due (sum of Boxes 1 and 2).
Box 4 VAT reclaimed in this period on purchases and other inputs (including acquisitions from other EU Member States).
Box 5 Net VAT to be paid to HMRC or reclaimed by you (difference between Boxes 3 and 4).
Box 6 Total value of sales and all other outputs excluding any VAT. Include the Box 8 figure.
Box 7 Total value of purchases and all other inputs excluding any VAT. Include the Box 9 figure.
Box 8 Total value of all supplies of goods and related services, excluding VAT, to other EU Member States.
Box 9 Total value of all acquisitions of goods and related services, excluding VAT, from other EU Member States.

The VAT Account

Registered businesses must keep a VAT account in permanent form as a record of the totals of the tax return for each tax period, together with the total of any tax payments or repayments.

VAT on Sales to a Trader in Another Country within the European Union

Each EU country has its own VAT rules and regulations; it is a complex area and it is wise to check these before embarking on overseas transactions. For more information see www.gov.uk.

Where goods are supplied to another EU country, they are technically known as despatches or removals rather than exports. The term 'exports' is reserved to describe sales to a country outside the EU. VAT may be charged on exports, see below. This applies even if you're sending them to another part of your own organisation. Where goods are supplied to a VAT registered business within the EU, the supply is zero-rated for VAT purposes, provided the following conditions are met:

- The goods are sent out of the UK to somewhere in another EU country.
- The recipient of the goods is genuinely registered for VAT.
- The VAT registration number of the EU trader – including the two letter country code – is shown on the sales invoice.
- Evidence that the goods have been despatched (usually within three months) to the EU trader is obtained.

To account for the VAT on zero-rated sales to another EU country, include the value of the goods and services on the VAT return in Box 6 and Box 8 in the usual way. If VAT is due in the destination country, the customer pays it to the tax office in their country.

All UK registered traders must send HMRC lists of their EU sales. There are three different forms required by HMRC regarding sales to traders in the EU:

1. The VAT return, Boxes 6 and 8.
2. The EU Sales List (ESL) automatically generated when Box 8 is completed on the VAT return.
3. The Intrastat Supplementary Declaration, which is for sales valued above the current limit for goods to other EU customers in a year.

VAT on Sales to a Trader Outside the EU (Exports)

A supply of goods sent to a destination outside the EU is zero-rated as a direct export where the following criteria are met:

- The goods are exported from the EU within the specified time limits (normally three months).
- Official or commercial evidence of export as appropriate within the specified time limits is obtained.
- Supplementary evidence of the export transaction is retained.
- The laws and the applicable conditions are complied with.

The supply must not be zero-rated as a direct export where:

- The customer's address is in the EU (including the UK).
- The goods are to be collected by or on behalf of the customer even if it is claimed they are for subsequent export.

Where the supply of goods to a trader for export is made via a third party in the UK, who is also making a taxable supply of goods or services to the same trader, the supply can be zero rated provided:

- The goods are only being delivered and *not supplied* to the third party in the UK.
- No use is made of the goods other than for processing or incorporation into other goods for export.
- The goods are exported from the EU and you obtain evidence of export within the specified time limits.

A record must be made of the:

- Name and address of the overseas trader
- Invoice number and date
- Description, quantity and value of the goods
- Name and address of the third party in the UK to whom the goods were delivered
- Date by which the goods must be exported and proof of export obtained
- Date of actual exportation.

VAT on Purchases from a Trader Outside the EU

Most services provided to business customers will be treated as supplied in the country where the business customer is established, and the business customer will account for VAT under the reverse charge mechanism. Services provided to non-business customers will still generally be liable to VAT in the country of the supplier.

To enable tax authorities to check that VAT is being accounted for correctly by the business receiving intra-EU supplies or services, UK VAT-registered businesses that supply services to EU businesses, where the place of supply is the customer's country, will have to complete the EU Sales List (ESL) for each calendar quarter and submit these within 14 days for paper returns and 21 days for electronic returns (in a similar way to someone who is supplying goods to other EU businesses).

The reverse of these rules is true for purchases of goods and services from overseas suppliers. For purchases from EU suppliers, reverse-charge rules are applied. For purchases from suppliers outside the EU, in most cases the invoice will be zero-rated.

Reverse-Charge Rules

The VAT account is credited with an amount of output tax, calculated on the full value of the supply received; at the same time, the VAT account is debited with the same amount of input tax to which you are entitled, in accordance with the normal rules.

This is shown in the VAT return as follows:

Box 2 Amount of output tax due on sales, EU acquisitions.
Box 4 Amount of input tax reclaimed on purchases.
Box 6 Full value of the supply.
Box 7 Full value of the supply, total value of purchases.
Box 9 Full value of the supply, value of EU purchases.

If the input tax is attributed to taxable supplies, and can therefore be reclaimed, the reverse charge results in no cost to the business. However, where this is not the case, VAT is payable on the supply at the UK rate, as if the supply was from a UK supplier.

Top Tips

- VAT is a highly complex area; never hesitate to check with the accountant or the helpline when in doubt.
- Where accounting software is used, always print the full range of reports that support the VAT return and keep a copy of the VAT return submitted.
- Farming businesses generally benefit from using the invoice method of accounting for VAT rather than cashbook.
- Check the tax point date on HP agreements, it may be later than the payment date and could fall into the following month for VAT purposes.
- Always file returns on time and ensure sufficient funds are in the bank account when a payment to HMRC is due by direct debit.

Chapter 8

Year-End Procedures

Having worked hard throughout the year to keep the bookkeeping records, it is crucial to go through some year-end checking procedures to identify any possible errors or omissions and prepare reconciliation schedules to support control accounts and demonstrate completeness. Detailed below are some routines and procedures that should be on your year-end (or month-end) accounts and bookkeeping checklist. Another annual task is preparing for the annual valuation; in this chapter we look at gathering physical information such as numbers of animals, tonnes of produce and quantifying consumables.

Checking for Errors or Omissions

Scan through the cashbook (or nominal ledger if computerised records are kept). Has everything been analysed correctly? It is easy to accidently put a figure in the wrong column or to miscode it. Highlight anything that looks out of place and reallocate it if appropriate.

- Confirm that the cashbook reconciles to the bank, see Chapters 4 and 6.
- Where differences arise, check for transposition errors. When two adjacent numbers are transposed, the resulting mathematical error will always be divisible by nine (e.g., (72–27) ÷ 9 = 5).
- Check that any recurring expenditure is recorded correctly; for example, are there 12 direct debits incurred during the financial year for business insurance? Are there 12 monthly rent receipts? If not, has the amount been misposted or does it need to be accrued?

- Compare the balances received and paid with the previous year. Are they consistent? If not, was this expected?
- Review the fixed asset register. Does the company still have all the fixed assets listed?
- Review the list of debtors. Are there any negative balances? This may indicate receipts on accounts that have not been matched to sales invoices. Are there any bad, or potentially bad, debts?
- Review the list of creditors. Are there any negative balances? This may indicate payments on account that have not been matched to purchase invoices. Are there any balances in dispute?
- Review private proportions. Have they been applied consistently? Have any costs been omitted from your calculations?
- Check payments received from the Rural Payments Agency, or other grant-awarding bodies, against agreements and claim forms.
- Check contract farming agreements/invoicing schedules – have interim invoices and divisible profit shares been settled by the year-end, or is it necessary to make provision for a debtor (contractor) or (creditor) farmer for services provided and divisible profits?

Accounting Reconciliations

Reconciliations are accounting processes used to compare two sets of records to ensure the figures agree and are accurate. For instance, do bank reconciliations prove the bookkeeping records to the bank statement? The reconciliation shows the balance on the bank statement and lists the adjustments necessary to get to the balance shown in the accounting records. The key accounting reconciliations needed are:

- Ensure the bank reconciliation is carried out. This is the most important reconciliation to ensure completeness.
- Prepare the VAT reconciliation. Does the balance owing or owed equal the closing VAT return? Can you explain any adjustments?
- Prepare the PAYE/NI reconciliation. Does the balance owing equal the balances on the payroll records?
- Check any business credit card statements to see that you have recorded all the business expenses in the books and records.

- Prepare a separate reconciliation statement for all business credit cards.
- Prepare a reconciliation of any loan accounts and hire purchase accounts. Do they agree to the statements provided?

Gathering Information for the Annual Valuation

The physical attributes of the farm are constantly changing; crops and animals are growing and being sold, materials are being purchased and used up, machinery is wearing out, buildings becoming in need of repair, or outmoded. In reviewing the activities of the business during the course of a financial year, it is clearly not sufficient merely to take account of the money flowing into and out of the bank account. Another complicating factor is the variability of seasons as they affect the farm.

It is therefore necessary when working out the farm profit (or loss) to take into account not only sales and purchases, but also changes in valuation. This means that an annual valuation must take place at the end of each financial year, one year's end-of-year valuation also serving as the following year's start-of-year figure.

Even where a business uses the services of a professional valuer for this annual task, preparatory work is needed, taking stock of everything owned by the business that makes up the valuation at the year-end. An established business is likely to have an inventory from the previous year that can be used for guidance. The various categories that make up the normal valuation and the information that will be needed for the valuation are as shown below. More information on the actual valuing process can be found in the next chapter.

Saleable Crops in Store

At the financial year-end, there may be a quantity of harvested crops that have yet to be sold. It is good practice to produce a reconciliation of harvested crops with assistance from the farmer. When harvest is complete, harvested areas and yields should be recorded. Self-bill invoices should be checked against crop contracts and an up-to-date store position maintained throughout the year. Where a sale has been agreed but the crop is still in store, it should be included in debtors at the full market price. However, there may also be a quantity of crops in store where no sale has been agreed. These need to be included in the farm valuation. The store may be on the farm itself or storage facilities of neighbouring farm, or a cooperative may be used. Where the crops are held in a pool by a cooperative, they should provide a valuation of the crops in store as at the financial year-end. Any pool payments received in advance of the final settlement should be treated as a loan. If crops are stored on the farm and accurate weights are unknown, estimates must be made (see Appendix 5). It should be possible to check the approximate quantity in store if an estimated harvested quantity is available minus any subsequent sales.

Growing Crops

The value of any growing crops should also be included in the valuation. This is arrived at by calculating the cost of the inputs (variable costs – seed, fertiliser sprays, etc.) and cultivations to the crop as at the year-end. Where a cropping program is used, it should be possible to produce a growing crops valuation. Alternatively, produce a schedule showing the cost of the inputs and any work carried out to the crop (ploughing, drilling, etc.).

Fodder and Bedding

Count bales of hay, silage and straw in store or, where the crop is in the ground, list inputs and cultivations as per growing crops.

Stores of Purchased Consumables

Make a list of stores of seed, fertiliser, sprays, feedstuffs, etc., fuel, oil, wearing machinery parts, building and fencing materials. Quantities and unit costs from purchase invoices will also be needed.

Machinery and Equipment

See 'asset register' (below).

Livestock

Check physical records (herd/flock) against accounting records at the end of the financial year. Where accounts are computerised and numbers of animals bought and sold have been recorded, it should be possible to produce reports showing annual totals of animals bought and sold for comparison with flock or herd records. In a cashbook system, a summary of the number of animals bought and sold can be made from notes made in the comments column or with reference to purchase or self-billing invoices. It is important to use annual totals adjusted for opening and closing debtors and creditors. For example, if 20 lambs were sold a week prior to the year-end and payment was not received for two weeks, the income would fall into the new accounting year, so the 20 lambs need to be bought into the correct accounting period (debtor).

Process for Calculating Livestock Numbers from Physical Records

- The number of animals recorded at the end of the previous financial year in the valuation and in the flock/herd record
- *plus* number of animals purchased, recorded in the accounts
- *less* opening creditors (number of animals) plus closing creditors (number of animals)
- *plus* births (physical records)
- *less* animals sold, recorded in the accounts; less opening debtors (number of animals) plus closing debtors (number of animals)
- *less* deaths
- = number of animals on the holding at the end of the current financial year, as per flock/herd records.

Updating the Fixed Asset Register

This is often a job left for the accountant but, with the aid of good accounting software, the process of keeping an asset register up to date is automated. However, where a cashbook system is used, it is possible to set up a spreadsheet that can be updated at the end of the financial year with dates, details and values of additions and disposals made during the year. See example below.

Depreciation

There are different methods of depreciation. The annual accounts of the business will state methods used and the rate at which certain asset classes are depreciated, in order to write off the cost less estimated residual value of each asset over its expected useful life.

Two common methods of depreciation are:

- **Straight line depreciation:** This is the simplest method, generally used for buildings. The total cost of an asset is divided by an estimate of its working life. The calculation should also have to take into account any selling or scrap value that might exist were the asset to be sold at the end of that working life, for example an asset worth £100,000 depreciated over ten years would be £10,000 per year.
- **Reducing balance depreciation:** This method is used for plant and machinery to calculate depreciation by reducing the value of the asset by the same percentage rate each year, therefore the amount of depreciation is greater in earlier than in later years, for example, £10,000 at 20%:

£10,000 – 20% = £2,000 depreciation in year 1 and a reduced balance of £8,000 – 20% = £1,600 in year 2 and so on.

HMRC sets rates of 'depreciation' that it is prepared to accept when calculating taxable profits; these rates are known as capital allowances. Different rates apply, depending on the type of asset, these are subject to review by the government in the Budget, up-to-date information can be found on the HMRC website. More general information on capital allowances is available at www.gov.uk.

Depreciation Calculated Annually or Monthly

Accounting programmes usually calculate depreciation monthly in order to produce financial management reports, this is summarised in a total at the end of the year, which gives the depreciation charge to be included in the accounts. Depreciation on additions and disposals are calculated from the date of purchase or sale. Where a cashbook is used, depreciation is usually calculated annually as shown in Figure 8.1. This will differ from the

Fixed Asset Register For the year ended xx/xx/x1																
	Purchase		Cost					Depreciation					NBV	NBV	Sale	Profit
	Date	HP	Bf	Additions	Disposals	Cf		Bf	Charge	On Disp	Cf		31/03/20x1	3 /03/20x0	Proceeds	(Loss)
Agricultural Building *10% Straight line*																
Original Building	xx/xx/xx		14,869.00	-	-	14,869.00		8,921.40	1,486.90		10,408.30		4,460.70	5,947.60	-	-
Plant & Machinery *20% Reducing Balance*																
Lawn mower	xx/xx/xx		425.00	-	-	425.00		395.00	6.00	-	401.00		24.00	30.00	-	-
Heater	xx/xx/xx		909.00	-	-	909.00		847.00	12.00	-	859.00		50.00	62.00	-	-
Electrical	xx/xx/xx		226.00	-	-	226.00		211.00	3.00	-	214.00		12.00	15.00	-	-
Hand Pallet Truck	xx/xx/xx		219.00	-	-	219.00		195.00	5.00		200.00		19.00	24.00	-	-
Robin BP253	xx/xx/xx		362.00	-	-	362.00		304.00	12.00		316.00		46.00	58.00		
Boiler	xx/xx/xx		1,330.00		-	1,330.00		894.00	87.00		981.00		349.00	436.00		
Mower	xx/xx/xx		574.00		-	574.00		386.00	38.00		424.00		150.00	188.00		
Strimmer	xx/xx/xx		208.00	-	-	208.00		75.00	27.00	-	102.00		106.00	133.00	-	-
Chain Saw	xx/xx/xx		250.00	-	250.00	-		50.00	-	50.00	-		-	200.00	100.00	(100.00)
Pick-up	xx/xx/xx		900.00	-	900.00	-		180.00	-	180.00	-		-	720.00	800.00	80.00
Brushcutter	xx/xx/x1		-	400.00	-	400.00		-	80.00	-	80.00		320.00	-	-	-
Bailey Trailer	xx/xx/x1	a	-	10,386.00	-	10,386.00		-	2,077.00	-	2,077.00		8,309.00	-	-	-
Landrover	xx/xx/x1		-	4,000.00	-	4,000.00		-	800.00	-	800.00		3,200.00	-	-	-
			5,403.00	14,786.00	1,150.00	19,039.00	-	3,537.00	3,147.00	230.00	6,454.00	-	12,585.00	1,866.00	900.00	(20.00)
TOTAL			20,272.00	14,786.00	1,150.00	33,908.00		12,458.40	4,633.90	230.00	16,862.30		17,045.70	7,813.60	900.00	(20.00)

10% of cost written off each year

20% of the remaining balance (cost less depreciation to date) written off each year

Enter value to reduce depreciation to nil

Enter value to reduce cost to nil

Enter the amount received

Enter the amount paid for new purchases

Figure 8.1 Fixed asset register

monthly calculation, since a full year's depreciation will be charged in the year of purchase of additions, even if the asset is bought at the end of the financial year, and no depreciation is charged in the year of sale, irrespective of the date of disposal.

Layout of the Fixed Asset Register

It is necessary for individual net book values (NBV) to be tracked for any asset. This allows you to calculate the profit or loss on the disposal of an asset. Companies also have to disclose the original cost and the depreciation of each class of asset in their accounts. The above layout provides the maximum amount of information that may be required in order to complete either a set of sole trader/partnership accounts or a full set of statutory accounts for a company.

The Year-End Checklist and File

Working with the accountant to prepare the accounts is an important part of the farm administrator's role. Talking with them to identify what information is required can save time and fees. Many schedules normally prepared by the accountant's junior staff could be prepared by the farm administrator instead, but it is important that this is agreed with the accountant beforehand. In addition, pulling together the key information that the accountant will need to see in an organised manner is helpful.

Below is a list of potential information that may be needed.

Bank

- Bank reconciliation/s at the year-end.
- Business bank statements for the year, plus month one of the following year.
- Any other bank building society statements.

VAT and Accounting Records

- Copy of cash analysis sheets and/or copy of Excel file for the accounting year (manual systems), data file plus printouts as agreed with accountant (computerised).
- Annual summary sheets for income and expenditure.

- Copies of VAT returns submitted in the year.
- Copies of any partial exemption calculations in the period, if applicable.
- Petty cash book and receipts, if applicable.
- Details of any HMRC inspection.

Debtors

- List any debtors at the year-end for manual bookkeeping or provide a copy of aged debtors' report if computerised.
- List any bad debts at the year-end.
- Copy invoices relating to prepayments.

Creditors

- List any creditors payable at the year-end for manual bookkeeping or provide a copy of the aged debtors' report if computerised.
- Copies of any bank loans and company credit card statements.
- Copy invoices relating to accruals.

Expenses

- File(s) of all invoices paid during the year.
- Copies of legal and professional fees invoices.
- Notes referring to any major fluctuations.
- Private proportion adjustments or percentage splits.

Income

- File(s) of all remittances and receipts for the year.
- Copies of miscellaneous income receipts and details as necessary.

Wages

- Summaries of wages and deductions for each employee for the year.
- Summary of payments made to HMRC.
- Details of any HMRC inspections.

Stock

- Schedule and notes for the valuation as detailed above.
- Delivery notes pre- and post-year end.
- Statement from merchants/co-operatives for crops held in pools at year-end.

Fixed Assets

- The fixed asset register (where maintained in house).
- Copy invoices of all additions in the year.
- Copy invoice of any disposals in the year.
- Copy of any new hire purchase agreements.
- Copy of any completion statements for sales of property.
- Notes regarding any items scrapped in the year.

Other

- Dividend statements for shares owned by the business.
- Copies of any new contract farming agreements.
- Copies of agreements, schedules and statements relating to government support payments and other grants.

Register of People with Significant Control

From June 2016, the Confirmation Statement replaces the Annual Return made by limited companies. A Register of People with Significant Control must be maintained and confirmed online with Companies House. Penalties apply if deadlines are missed.

Top Tips

- Keep stock-taking inventories and calculations in a spreadsheet; this will act as a starting point for the following year.
- Make use of the asset register in accounting software; time taken to set it up is worth it in the long run.
- Check that the opening trial balance has been adjusted, as per advice from accountant (if this is carried out in house), before producing reports and running the year-end.
- The year-end is a good time to make improvements to the accounting system; for example, using a spreadsheet instead of a manual cashbook or considering investing in an accounting program.
- Follow the instructions from the software provider for opening the new accounting year and closing down the old one.

Chapter 9

Profit and Loss Account

Bookkeeping provides the details of a business's transactions for a prescribed period, but this is insufficient in determining the profitability of the business. Profit may be hidden in an increased value of stock held or the value of products used for personal use. The yearly transactions, therefore, have to be adjusted for various factors to give a true picture of how well the business has performed. The purpose of the profit and loss account is to show the true business results over a period of time. The profit and loss account forms part of the financial statements that most businesses prepare at the end of the year. These are used for various purposes including:

- To inform HMRC about the profits the company has made on which they will need to pay tax.
- To inform various stakeholders about the performance of the business over the period. This may include shareholders, investors and banks.
- As a permanent financial history of the business. Many businesses will refer back to their performance in previous years to judge how well they are doing in the current period or to budget for future periods.
- As proof of financial well-being, when applying for credit for a loan or a hire purchase agreement.

Generally, farm offices have systems in place to keep the cashbook records, either manually or using a computer system; however, many do not use this information to go on to prepare a profit and loss account. It can be very useful to know what is needed to prepare the profit and loss account, so that you can:

- Ensure that all the information that the accountant might ask for is available.
- Estimate the tax that will be payable based on the draft profits and therefore budget for the future cost.
- Base management decisions on the true performance of the business.
- Be able to benchmark the business against other similar businesses, past results or industry statistics.

The profit and loss account will attempt as far as possible to match the expenses for a period against the income to which it relates. Normally for farming businesses, the income will reflect the harvest that falls within the year or the livestock sales that occur within that period. The costs that relate to the production of those crops and livestock that have been sold will be the expenses that need to be included in the accounts. In addition, there will be a variety of overhead costs that will be included based on the period to which they relate. The adjustments to the bookkeeping records that might be needed are therefore as follows.

Sales (Otherwise Referred to as Gross Output)

Arable Sales

The accounts will show the income from the sale of crops relating to the harvest that falls within the accounting period. For example, for an accounting period of 31 March 20x1, all sales relating to crops harvested between 1 April 20x0 and 31 March 20x1 will be included. This may include those shown in Diagram 9.1.

There may be some crops that were harvested in August 20x9, but which

Crop	Harvested
Barley	June 20x0
Oil Seed Rape	July 20x0
Wheat	August 20x0
Beans	August 20x0
Peas	August 20x0
Sugarbeet	September – December 20x0
Potatoes	September 20x0

Diagram 9.1

were held in store and sold in April 20x0. Because these were harvested in the year to 31 March 20x0, they should be recognised as sales for this year and not as sales within the current year to 31 March 20x1. This is achieved by including a value for the crops in the closing stock valuation as at 31 March 20x0. The double entry for this will increase closing stock on the balance sheet (debit) and increase sales in the profit and loss account (credit). The sale is not included in the accounts at the full sale value, however, because the sale has not yet taken place and you should only recognise profit when it has actually arisen. The closing stock is therefore valued at the lower of the cost of production or the net realisable value. HMRC has agreed that an acceptable method of estimating the cost of production is to use 75% of the market value. More information can be found at www.hmrc.gov.uk, Help sheet 232.

For the accounts to 31 March 20x1, this stock then reverses through as opening stock – which will be shown as a cost in the profit and loss account. When the crops in opening stock are eventually sold, the sale is therefore matched against the opening stock value, leaving just the profit element showing in the accounts for the current year. There may then be a similar adjustment for closing stock in the 20x1 accounts.

Livestock Sales

Livestock sales and the sales of goods produced by livestock, including milk, eggs and wool, are included in the accounts for the year in which the sale arises. A sale will have arisen when it is unconditional, which may not necessarily be when the payment is received.

As a normal rule, farm animals are dealt with as trading stock, and therefore sales are adjusted as with arable crops for movements in stock. As mentioned above, stock is valued at the lower of cost and net realisable value, and for livestock HMRC will accept a deemed cost calculation as follows:

- Cattle – 60% of open market value.
- Sheep and pigs – 75% of open market value.

Some farm animals that are kept not primarily for resale but for their products (e.g., milk, eggs, or wool) or offspring (e.g., lambs or piglets) are more like capital assets than trading stock. Tax law recognises this by giving farmers the option of accounting for such 'production animals' under the herd basis. A farmer must elect for the herd basis, the election must specify the class of animals concerned and, once it has been made, the herd basis must be used for as long as the farmer continues to keep animals of that class. The election has to be made soon after the farmer first starts keeping animals of that class, and then the herd basis applies to those animals from the outset. There is a further opportunity to elect to use the herd basis if a herd or a substantial part of it (20% or more) is compulsorily slaughtered. The accountant is likely to make the necessary adjustments if the farm has elected to use the herd basis, but if an election has been made, then it will affect the way that any disposals and purchases of those animals are accounted for. Further information can be found at www.hmrc.gov.uk, Help Sheet 224.

Variable Costs

Variable costs are the expenses that a farming business incurs that change in proportion to the activity of a business.

Arable Variable Costs

The cost of seeds, fertilisers, sprays and some contracting expenses will vary depending on the amount and type of crops grown. The profit and loss account should include those expenses that relate to the harvest that falls within the accounting period. In the 31 March 20x1 example given above, the wheat sales that are included in the profit and loss account are all those that relate to the crop harvested in August 20x0. Therefore, the costs relating to growing that crop should also be included in that profit and loss account. Any costs relating to the following harvest should be carried forward in closing stock.

Livestock Variable Costs

Similarly, the direct costs relating to the rearing of the animals sold, or the production of the goods from those animals, should be included in the same profit and loss account as the income from those animals. Typical examples of livestock variable costs are concentrates, forage costs, haulage and animal health expenses.

Gross Margin

The gross margin of an enterprise can be defined as the income (or the gross output) less its variable costs. Most farm accounts that are produced by accountants that specialise in preparing farm accounts will produce the gross margin for each individual enterprise. This may involve preparing separate trading accounts for each enterprise for larger businesses, or for smaller businesses it may be part of one trading and profit and loss account. This assists the reader of the accounts to identify how individual parts of the business are performing so that each can be judged accordingly.

Other Income

Government Support Payments

The general principle for including government grants or subsidies within a set of accounts is that:

- the business should be reasonably sure that it will comply with the conditions attaching to the grant/subsidy; and
- the grant/subsidy must be included in the same period as any relevant expenses, i.e., it should follow normal accruals principles.

The Basic Payment Scheme (BPS) replaced the Single Payment Scheme (SPS) on 1 January 2015, claimants now only need to have the land used to match their entitlements at their disposal on one day (which, for the UK, has been set as 15 May) in each scheme year. In addition, such land must also be eligible for the entire calendar year. Further information can be found at www.gov.uk.

Rental Income

The profit and loss account should include all rents that are receivable for the relevant period, whether or not they have actually been received. Therefore, the profit and loss account will need to accrue for any rents that are due, but that have not yet been paid by the business's tenants by the end of the period. Similarly, if any rents have been paid in advance, the profit and loss account will need to be adjusted to defer the proportion of income that relates to the following period. Where rents are relatively small, HMRC will accept accounting for rental income as part of the overall trading income.

Contracting Income

Some farming businesses will also carry out contracting work for other farming businesses. HMRC does not regard this income as farming income, but it is trading income. This is important when considering farming-specific tax reliefs such as averaging. The profit and loss account should include the income for all work carried out as at the year-end date. Any sales that have been invoiced but not received will be included as a debtor, whereas any work that has not yet been invoiced should be included as part of work in progress in the closing valuation.

Overhead Costs (Also Referred to as Fixed Costs)

Overhead costs are expenses that, although incurred in a business's day-to-day operations, cannot be directly assigned to a specific enterprise. These are sometimes organised under the following group headings.

Labour

- Wages and related PAYE and NI costs
- Pension contributions
- Training expenses.

Power and Machinery

- Fuel and oil
- Machinery repairs
- Licence and insurance expenses
- Machinery depreciation.

Property Costs

- Rates
- Rent
- Light and heat
- Property repairs

- Insurance
- Property depreciation.

Administration

- Business insurance
- Telephone and office expenses
- Legal and professional fees
- Subscriptions and donations
- Depreciation on fixtures and fittings.

Finance

- Bank and loan interest
- HP interest.

All of the above costs should be adjusted on an accruals basis, so that the costs cover the full period of the accounts. The most common areas to review are:

PAYE/NI: It is common for the final month's PAYE/NI to be outstanding, as it is usually paid 19 days after the month-end.
Rent: Ensure that the rent paid is only for the period of the accounts. If rent is paid in advance, a prepayment may be required to reduce the costs included in the profit and loss account. If it is paid in arrears, then an accrual may be needed for the additional rent.
Insurance: Insurance payments are often paid for the full year, but this rarely coincides with the year-end. A prepayment or an accrual may be necessary depending on when the insurance period ends in relation to the year-end.
Telephone and utility bills: These are often paid quarterly in arrears, and this may not coincide with the year-end. An accrual may be required to include a full year's costs.
Subscriptions: Subscriptions such as the NFU or CLA membership subscription are often paid on an annual basis. A prepayment or accrual may therefore be required to ensure a full year's costs are allocated in the accounts.
Private proportions: If these have not already been identified and accounted for within the bookkeeping records, an adjustment will need

to be made for any private element of the overhead costs. For example, a share of the utility bills for the farmhouse will need to be deducted as they relate to the personal living costs of the farmer rather than the trade.

Net Profit

The gross margin, plus other income, less overhead costs results in the farm net profit for the year.

Profit and Loss Account Format

The format of the profit and loss account will depend largely on the preferences of the accountant who prepares the accounts, and those of the farming client. For sole traders and partnerships there is no required format, but companies will need to prepare a statutory profit and loss account. Companies may also wish to prepare a detailed profit and loss account in a similar format to that prepared by partnership and sole traders. An example of this is shown below. A statutory profit and loss account will include different layout and terminology to describe the income and expenditure, but of course the principles for calculating those figures are the same. Some of the different terms used are:

Turnover: Total sales income that a company receives from its normal business activities, usually from the sale of goods and services to customers. This generally equals the farm gross output, but may also include some other income, such as contracting.

Cost of goods sold (COGS): All costs incurred in purchasing or producing the goods and services included in turnover. This generally equals the variable costs.

Other operating income: Income from activities other than normal business operations, such as rental income or profit from the sale of assets.

An example of the layout of the detailed profit and loss account is shown in Diagram 9.2.

A Farmer & Sons

Profit and Loss Account
For the Year ended 31 March 20x1

	Page	£	20x0 £
Gross Margin			
Livestock Trading Account	3		53,037
Arable Trading Account	4		17,213
Farm Gross Margin			70,250
Other Income			
Rental income		950	
Contract work		650	
Single Farm Payment		9,000	
			10,600
Gross Profit			80,850
Less Overhead Costs			
Labour		14,125	
Power and Machinery		15,923	
Property Costs		9,034	
Administration		13,029	
Finance		3,547	
			55,658
Net Farm Profit			25,192

Figure 9.2 Profit and loss account

A Farmer & Sons
Livestock Trading Account
For the Year ended 31 March 20x1

		20x1
	£	**£**
Livestock sales		
Sale of animals		14,800
Transfer to herd		1,200
Milk		68,337
Gross Output		84,337
Variable Costs		
Opening valuation	32,500	
Lease of milk quota	4,620	
Haulage	365	
Veterinary expenses	1,215	
Feedstuffs	12,800	
	51,500	
Closing valuation	20,200	
		31,300
Gross Margin		53,037

A Farmer & Sons
Arable Trading Account
For the Year ended 31 March 20x1

		20x1
	£	**£**
Arable sales		
Wheat		15,063
Barley		8,100
Oilseed Rape		10,000
Straw		50
Gross Output		33,213
Variable Costs		
Opening valuation	10,000	
Seed, fertiliser & Sprays	30,000	
	40,000	
Closing valuation	24,000	
		16,000
Gross Margin		17,213

Diagram 9.3

A Farmer & Sons
Overhead Costs
For the Year ended 31 March 20x1

		20x1
	£	£
Labour		
Wages		14,125
Power and Machinery		
Machinery & vehicle expenses		15,125
Property Costs		
Rent		2,350
Rates		1,700
Insurance		1,205
Light & heat		3,779
		9,034
Administration		
Telephone		759
Contracting and Plant hire		1,991
Silage wrapping		1,592
Accountancy		923
Professional fees		175
Miscellaneous expenses		1,297
Depreciation		6,292
		13,029
Finance		
Bank charges		2,273
Hire purchase charges		1,274
		3,547

Diagram 9.4

Top Tips

- The profit and loss account reflects the success or otherwise of a business over a period.
- If data is dealt with as it arrives in the office, and kept up-to-date, it should be possible to produce an estimated profit forecast in time to make reasonably informed decisions in advance of the close of year, which may affect tax liability.
- Aim to finalise accounts about a month after the year-end, allowing time for all purchase and sales invoices to be processed and statements to be agreed with account balances.
- Make use of the annual accounts, research any large variations, year on year.
- Don't confuse profit with cash flow.

Chapter 10

Balance Sheet

The balance sheet is one of the most important statements in a company's accounts. It shows what assets and liabilities a company has, and how the business is funded (by shareholders/proprietors and by debt). In contrast to the profit and loss account, which reflects the success or otherwise of a business over a period, the balance sheet is a self-diagnosis at any one point in time. It is a snapshot of how healthy the business is at that exact point. The balance sheet is prepared as part of the business's financial statements at the end of each financial year. The opening balance sheet – together with the profit and loss account for the period and the closing balance sheet – therefore, gives the full picture of how the business was to start with, how it got on during the year and how it is at the moment. Thus, the three documents together have considerable diagnostic value, not only to the farmer but also to the farm's advisers, bank managers and other current or potential creditors. The layout of the balance sheet tends to follow a set pattern, showing the assets of the business first then deducting the business's liabilities to leave the net assets. To be able to prepare the balance sheet, an understanding of the separate components that make it up and how they are calculated is necessary, these are as follows.

Fixed Assets

Fixed assets, as opposed to current assets, are those assets with a remaining useful life of more than one year. These assets are shown on the balance sheet but their value is depreciated, and the depreciation is treated as an expense in the profit and loss account for each year of their life.

There are two types of fixed assets:

- Tangible fixed assets.
- Intangible fixed assets.

Tangible fixed assets include physical assets, such as land and buildings and equipment. Long-term financial investments are also considered tangible.

Intangible assets are assets that you cannot physically hold, but that still add value to the business. The most important intangible fixed asset is goodwill. Other intangibles include patents, copyrights and trademarks.

Although the commonest way of accounting for the limited useful life of a fixed asset is to depreciate it, intangible assets are amortised (essentially the same) and assets whose use can be measured (such as mines) may be depleted.

Assets that do not have a limited life or that keep their value (such as land and investments) may not need to be depreciated, amortised or depleted. It is normal for the chosen method of depreciation to be written in the notes to the accounts.

Current Assets

Current assets are those assets that are expected to be used (sold or consumed) within a year, unlike fixed assets. Current assets are shown on the balance sheet, and are listed in order of increasing liquidity (i.e., how easy they are to convert to cash). Usually stocks (of goods) will be listed first, followed by debtors, with cash last.

The current asset position of a business is important, both for assessing its financial strength and for gauging its operational efficiency.

Closing Valuation (Stock)

Stocks are current assets held for sale, or for processing and subsequent resale. Stocks should be valued at the lower of:

- Cost (purchase price + the cost of any processing).
- Net realisable value (i.e. what its potential market value is).

Stocks may consist of:

- Finished goods (e.g., corn in the store ready for sale).
- Work in progress (e.g., seeds, fertilisers and sprays applied to next year's harvest).
- Raw materials (e.g., fuel or fertilisers in-store).

Debtors

Debtors represent income that is due to the business. There are various types of debtor that may be included as follows:

Trade debtors: Amounts owed by customers, where the business has raised an invoice for the sale of goods or services, but this has not been settled at the balance sheet date. This may also be referred to as the sales ledger, debtors' ledger or debtors control account.

Prepayments and accrued income: Prepayments are expenses paid for by the business that relate to the following year. For example, a farm with a 31 March year-end pays the business insurance of £500, which covers the year to 30 June. Nine months of the premium relates to the year to 31 March, but the other three months relates to the following year. A prepayment of £125 (3/12 × £500) would need to be provided. This would reduce the insurance cost shown in the profit and loss account and increase the debtors shown on the balance sheet.

Accrued income is the income that has been earned by the business by providing a service or selling a product, but has yet to be received. For example, assume that the farm lets a cottage for six months, but under the terms of the lease, the rent will be paid in full at the end of the six months. If the year-end falls in the middle of that six months, the farm will need to accrue for the rent that is due to it up to the financial year-end, even though physical payment is not received until the end of the six months.

Other debtors: Other typical debtors are the VAT repayment due for the final month/quarter, the balance of any amounts advanced to a director (but not a partner or sole trader), a corporation tax refund or the sale of an asset – where the transaction has taken place but no money has been received by the year-end.

Cash at Bank and In-Hand

This usually includes bank accounts that are in credit and any cash that is held. The cash at bank is not the balance held in the bank account, but the balance adjusted for items that have not cleared the bank, as discussed in bank reconciliation in Chapters 4 and 6.

Current Liabilities

Liabilities that are to be settled in less than a year are called current liabilities. These include:

Trade creditors: These are the business's liabilities to its suppliers. This may also be referred to as the purchase or creditors' ledger or the creditors control account.
Bank overdraft: Where the bank is no longer in credit, it cannot be shown as an asset under cash at bank and in-hand. Since an overdraft may be withdrawn without notice by a bank, it is included under current liabilities.
Bank loans: These may be repayable over a number of years. That part of the loan that is payable within one year should be included under current liabilities.
Hire purchase: The capital element of any hire purchase (i.e., excluding interest charged) should be included as a creditor. The part of that balance that is repayable within one year should be included under current liabilities.
Accruals and deferred income: Accruals are expenses that are recognised in the accounts before they are physically paid. They are recognised because of the extremely high likelihood of payment. Accruals are generally periodic payments. For example, in order that the accounts can be completed they will need to be prepared by the accountant. The fee for this is not normally paid until after the year-end; however, this cost directly relates to the year in question and an accrual is therefore made to include it in the accounts.

Deferred income is income that has been received in the financial year, but that actually relates to a future period. For example, the farm may let a cottage for six months but under the terms of this lease, the rent is to be paid in full at the beginning of the six months. As the year-end falls in the middle of that six months, the farm will need to defer three months of the rent as it relates to the following financial period.

Tax: At the year-end, there is likely to be a number of taxes owed. This may be the closing VAT balance if money is owed to HMRC rather than due from them, corporation tax may be payable for companies and PAYE/NI may be due on any salaries paid in the final period. Companies may also have an adjustment for deferred tax (sometimes referred to as 'provision for liabilities'). Deferred tax is an accounting adjustment. Deferred tax liabilities are provided in order that investors may understand the future tax liabilities that may arise as a result of either:

- tax reliefs that have reduced current profits but which will result in higher profits at a later date, or
- where income included in the accounts has not yet been taxed.

These are referred to as timing differences.

It is also possible to have a deferred tax asset on the balance sheet. The main reason for this is where there are losses available to reduce future profits. Where there is a deferred tax asset, this is usually included within other debtors.

Other creditors: Other creditors include all other liabilities that will fall due within the next year, or are able to be called in without notice. This typically includes loans from directors.

Long-Term Liabilities

Long-term liabilities are those that are due to be paid in more than a year. Typical examples are long-term loans and HP agreements. Only the amount payable after more than one year should be included in long-term liabilities.

The balance sheet is split into two equal sections. The net assets shown in the top half of the balance sheet will be equalled by the capital and current accounts for sole traders and partnerships, but for companies the net assets will equal the total of:

- Share capital
- Share premium account

- Other reserves (such as the revaluation reserve)
- Retained earnings (the sum of profits and losses from previous years to date).

An example of the layout of the balance sheet is shown in Diagram 10.1.

A Farmer & Sons
Balance Sheet
As at 31 March 20x1

	Notes	£	20x1 £
Fixed assets			
Tangible fixed assets	1		358,541
Intangible fixed assets	2		9,520
Dairy Herd	3		17,700
			385,761
Current assets			
Stock	4	44,200	
Debtors and prepayments	5	14,770	
		58,970	
Current liabilities			
Creditors	6	16,021	
Bank overdraft		11,565	
Hire purchase account		5,641	
		33,227	
Net Current assets			25,743
Total assets less current liabilities			411,504
Creditors due after more than one year			
Hire purchase account			(20,000)
Net assets			391,504
Represented by:			
Capital account			
Balance as at 1 April 20x0			384,623
Net profit			25,192
			409,815
Less drawings			18,311
			391,504

Diagram 10.1

Top Tips

- The balance sheet reflects the health of a business at a given date.
- A liability is considered current if it is due within a year. An asset is current if it can be converted into cash within a year.
- Fixed assets, as opposed to current assets, are those assets with a remaining useful life of more than one year.
- Current ratio = current assets/current liabilities, in the example above, 58,970/33,227 = 1.75, a high ratio indicates that a company can pay its creditors as they fall due. A number less than one indicates potential cash flow problems. See Appendix 3 for more KPIs (Key Performance Indicators).
- Read and understand the annual accounts and the implications of the results and trends.

Chapter 11

Management Reporting and Budgeting

The first chapter of this handbook touches on the need to produce an annual budget for the forthcoming year and the role played by the farm administrator. This chapter will look at it in more detail.

If no major changes are to be made, such as a new enterprise or major capital expenditure, overheads will remain in line with inflation and reference may be made to the preceding year's accounts as a starting point. However, the cropping plan, livestock numbers, market trends and timing of sales will have an effect on income and variable costs and this information is more likely to come from the farm manager.

It is important that the preparation of the annual budget be a team effort. It may be prudent to time the purchase of a major piece of farm equipment before the end of the tax year in order to take advantage of the capital allowances prevailing at the time, thus reducing imminent tax liability. Income from DEFRA's Rural Payments Agency sometimes forms a major part of farm income and needs careful consideration. A schedule of payments under existing agreements can be built into the budget, but are there any plans to apply for new schemes in the forthcoming year that need to be taken into account? Again, the manager is more likely to know about this than the administrator.

Management Reporting

This section looks at the basic principles of physical planning in order to budget for the future. The Farm Administrator should have a good understanding of each enterprise on the farm and work closely with managers.

It is advisable to start with a cropping plan and build up a picture of the seed, fertiliser and sprays required and the estimated yields.

An example of a cropping schedule with seed requirement is shown below:

Field Name/no.	ha	Cropping current year				Cropping following year	Seed Requirement	
		Wheat	OSR	Grazing	Silage		Wheat t. @ 165kg/ha	Grass t. @ 35kg/ha
High Wood	10.3	10.3				Grass – silage		0.36
Brick Kiln	8.0				8.0	Grass – silage		
Scrub	9.5			9.5		Grass – grazing		
High Barn	12.6		12.6			Winter Wheat 1	2.08	
Home Field	14.1				14.1	Grass – silage		
Rowdown	19.2	19.2				Winter Wheat 2	3.17	
Kings Field	4.0			4.0		Grass – grazing		
etc.								
etc.								
Total	77.7	29.5	12.6	13.5	22.1		5.3	0.4

Diagram 11.1

This can be adapted for fertiliser and spray usage.

It is important to separate the different crop years and then draw on the relevant information for a particular financial budgeting year. It may be that the financial year of the business does not coincide with the harvest year and therefore your cropping schedule should cover two years for ease of extracting information.

Livestock: The essential part of physical planning here is the livestock movement plan (see Diagram 11.2).

Having completed the movement plan, the opening and closing numbers can be used to calculate the livestock valuations, and sales and purchase numbers can be used for financial budgeting along with the feed requirements for the numbers on farm at any time. This movement plan is for a dairy herd but can be adapted to cover a suckler herd, sheep, pigs, poultry, etc.

The recordkeeping now required for cross-compliance, whether it is done manually, with a computer spreadsheet or specialised software programme,

Type of Stock	Dairy Cows	Calves	Bulls	Heifers <12mths	Bulling Heifers	Check Column
Movement Opening numbers	150	20	1	21	23	215
births		145				145
transfers in	23			20	21	64
purchases			1			1
deaths	-2					-2
transfers out		-20		-21	-23	-64
sales	-21	-120	-1			-142
Closing numbers	150	25	1	20	21	217

Diagram 11.2

makes storing and extracting the necessary information for budgeting considerably easier.

The farm administrator should also be aware of any expected overhead expenditure and keep a checklist of planned work in the following areas:

- Building repairs
- Fencing and road repairs.

It is also advisable to have a reminder of dates for insurance renewals, rent reviews, bank overdraft and anything else that may need forward negotiation to get the most favourable rates.
The basis of this can be brought forward from the previous year's figures with appropriate adjustment for inflation and changes.

Budgeting

It has been emphasised in earlier chapters that the purpose of management reporting is to provide information that can be used in making decisions for the future. It is not just something to do to keep the bank manager happy.

A budget is simply a forecast of effects; this is easy to prepare if the correct information is available. Whole farm budgets are necessary before taking over a farm and for planning for the future. The two most-used

Check list for overhead & capital planning			
Labour	Regular wages (gross)		
	Casual wages		
	Employers NI contributions		
		Total	£
Vehicles & Machinery	Machinery/vehicle repairs		
	Fuel		
	Vehicle taxes		
	Vehicle insurance		
	Contract & hire		
	M & V depreciation		
		Total	£
Power, light & heat	Electricity		
	Gas		
	Oil		
		Total	£
Administration	Office costs		
	Professional fees		
	Telephone		
	Insurance		
	Miscellaneous (subs. etc.)		
		Total	£
Property	Property repairs		
	Water & drainage		
	Other property costs		
	Property depreciation		
		Total	£
Rent & Finance	Rent		
	Bank Charges		
	Bank Interest		
	Other		
		Total	£
General Contract Requirements:	List as required		
		Total	£
Planned Capital Expenditure:	List as required		
	e.g. new pick-up truck		
		Total	£

Diagram 11.3

types of financial budgets are gross margin (income/profit) budgets for each enterprise and cash-flow budgets, which will also reflect capital requirements. However, one can be highly misleading without the other and it is advisable to reconcile these to each other.

The true state of the business can only be known if the cash flow is reconciled to a profit and loss account for the year, so it is important to have:

- An annual profit and loss forecast.
- A cash flow forecast (preferably monthly) for the period.
- Detailed physical and financial assumptions behind the forecasts; for example, area of crops, yield and prices, numbers of sheep, lambing percentage, numbers of cattle and sale values.

For the income/profit budget, the management team will need to work out the predicted gross margin for each enterprise for the forthcoming financial year.

Gross Margin Budget

The manager will be seeking answers to 'What would happen if…?' questions. These will mostly be relatively minor changes of emphasis between one enterprise and another. The need to budget for a revolution in the whole business structure of the farm can occur only rarely. Therefore, the data that is most likely to be needed are those concerning cropping gross margins per hectare and livestock gross margins per head. Such information can be derived from the cash analysis accounts on a whole-enterprise basis. Dividing by the area concerned or the number of animals involved converts this to a per-hectare or per-head basis.

Naturally there are pitfalls to be avoided. Perhaps the most obvious source of possible confusion concerns the distinction between financial years and farming years. Cereal seed may be purchased in year A for growing and harvested in year B, to be stored on the farm and eventually sold towards the end of year C. Thus, during any particular financial year, the variable cost items may be for next year's crops, while the returns relate to last year's crops; the current year's crops may hardly feature at all. Similarly, this year's beef enterprise variable costs feature alongside returns from animals fed last year.

In a static situation, these complications have no effect. But farming is never static. There are changes in prices, returns, yields, forage quality, techniques, personnel and probably in size of enterprise from year to year. It is therefore the job of the management team to produce meaningful budget data by relating cash analysis figures to the relevant physical data.

The gross margin = return (sales, and values of transfers to other enterprises) minus the variable costs.

It is useful to include any income from RPA and other environmental schemes under the bottom line of the gross margin so that it does not distort the profitability of the enterprise.

It must be emphasised that gross margin is not profit. Gross margin is purely and simply the contribution made by an enterprise towards covering the fixed costs of the business. The fixed costs have been defined in the management reporting checklist and will include any finance or loan repayments.

Annual profit and loss forecast = total gross margins minus total fixed costs.

Benchmarking

The most useful gross margin figures are those for next year. The latest available farm gross margins might be increased by an inflationary multiplier or enterprise industry standard information (benchmarking) can be used.

Benchmarking allows the comparison of the financial performance of a business to the performance of average and top performing farms of the same farm type. This system also allows comparison of gross margins, profit and loss account, balance sheet and performance measures. See the Rural Business Research (RBR) at www.farmbusinesssurvey.co.uk.

Commonly used performance measures, such as the calving index or the lambing percentage, are explained in Appendix 2.

Information such as 'normal' wheat yield per hectare, 'normal' number of piglets per sow per year, plus the farm's known normal inputs in terms of seed rates per hectare, kg/ha of N, P and K, etc. The initial setting up can be time consuming, but modification for use in subsequent years is relatively straightforward. The physical data are then multiplied by the manager's best estimates of next year's prices.

Compare enterprises within a farming business, year-on-year results and results to national benchmarks.

Gross Margin calculation based on an arable unit with a cropped area of 138 hectares

Gross Output

Crop	Has	X Tonnes/Ha	= Total Tonnes	X Price per Tonne	= Total Output
Winter Wheat	53	6	318	150	47700
Spring Barley	19	3	57	185	10545
Oilseed Rape	40	4	160	320	51200
Sugar Beet	26	55	1430	23	32890
	138				142335

Less Variable Costs per ha

Seed	+ Fertiliser	+ Crop Protection	+ Sundry (Levies)	+ Haulage	= Total VCs per Ha	X Has = Total VCs
60	252	155	16		483	25599
50	170	110	10		340	6460
68	232	148			448	17920
187	290	168	11	148	804	20891
						70870

= Gross Margin

Output-VCs total	per Ha
22101	417
4085	215
33280	832
11999	462
71465	518

Gross Margin based on a livestock unit with a grazing area of 77 ha, with 140 Dairy cows + followers and 300 lowland breeding ewes

Gross Output

Livestock	Numbers	X unit per Head	= Total Unit	X Price per unit	= Total Output
Dairy Cows a) milk income	140	7000	980000	0.25	245000
b) calf income	140	1	140	153	21420
c) less depreciation	140	1	140	-263	36820
					303240

Less Variable Costs per cow

Concentrates	+ Bulk Feed	+ Forage	+ Vet Med & AI	+ Other VCs	= Total VCs per Cow	X Numbers = Total VCs
334	50	83	80	84	631	88340

Output-VCs total	per Cow
214900	1535

Gross Output

Livestock	Numbers	X unit per Head	= Total Unit	X Price per unit	= Total Output
Lowland Ewes a) lamb income	300	1.65	495	85.00	42075
b) wool income	300	1	300	1.50	450
c) less depreciation	300	1	300	-20.00	-6000
					36525

Less Variable Costs per ewe

Concentrates Ewes	+ Concentrates Lambs	+ Forage	+ Vet Med & AI	+ Other VCs	= Total VCs per Cow	X Numbers = Total VCs
10	11	13	6	8	48	14400

Output-VCs total	per Ewe
22125	74

Diagram 11.4

Cash Flow Budget

The distinction between profit and cash flow has already been pointed out. The manager must satisfy himself not only that his plans will result in profits, but also that there will be sufficient cash or credit to put the plans into effect. This is particularly important in the early stages of creating a new business and in expanding an existing enterprise substantially. However, if credit facilities are involved, the bank manager will insist on seeing a cash flow budget; he must not only be assured of the safety of his loan money, but require a realistic indication of the ability to pay interest and the timing of repayments.

Again, the responsibility for the cash flow budget rests on the manager, but the farm administrator may well be involved, either in helping to prepare the budget or in compiling a cash flow record of the past 12 months, which the manager can then adapt for budget purposes. No general rules can be laid down since the cash flow information required depends on the use to which the budget will be put and the type of enterprise change being contemplated.

The most likely requirement, however, is that there shall be a month-by-month summary of cash coming into and going out from the farm account under a number of headings:

- Variable costs and returns of enterprises.
- Farming and environmental subsidies.
- Total of regular fixed costs including capital repayments already committed.
- Itemised list of exceptional fixed costs and capital purchases.
- Total of regular private costs.
- Itemised list of exceptional private items.

Spreadsheets are often used to make cash flow summaries and budgets, but most farming software has the reporting ability to use the inputted data to produce a cash flow and transfer it to a budget programme making this a relatively simple exercise.

Remember – only the past can be recorded, the future can only be planned, which emphasises that all accounts, all field records, all livestock records, etc. refer to times past but can give guidance to planning the future.

Year	CASHFLOW BUDGET										
	b/f	Month 1		Month 2		Month 3		Month 4...............		Cont..to Month 12	
	Actual	Budget	Actual	Budget	Actual	Budget	Actual	Budget	Actual	Budget	Actual
INCOME											
Winter Wheat											
Spring Barley											
Straw & Hay											
Sheep Sales											
Support Payment & Grants											
Contract & Hire											
Farm Shop Sales											
Sundry Income											
Capital Sales											
Private Income											
VAT Refund											
Fuel Scale Charge											
Vat Charged on Sales											
TOTAL		0.0	0.0	0.0	0.0	0.0	0.0	0.0	0.0	0.0	0.0
EXPENSES											
Seed											
Spray											
Fertiliser											
Crop Sundries											
Livestock Purchase											
Feed											
Vet & Med											
Stock Sundries											
Wages & PAYE											
Machinery Repairs											
Vehicle Expenses											
Fuel,Oil, Electric											
Contract, Hire & Transport											
Office & Prof Fees											
Rent, Rates, Water											
Property Repairs											
General Insurance											
General Sundries											
Finance Charges											
Capital											
Private Drawings											
VAT Payable to HMRC											
Fuel Scale Charge											
VAT to Reclaim											
TOTAL		0.0	0.0	0.0	0.0	0.0	0.0	0.0	0.0	0.0	0.0
Cash Flow for Month		0.0	0.0	0.0	0.0	0.0	0.0	0.0	0.0	0.0	0.0
Opening Balance		0.0	0.0	0.0	0.0	0.0	0.0	0.0	0.0	0.0	0.0
Closing Balance		0.0	0.0	0.0	0.0	0.0	0.0	0.0	0.0	0.0	0.0

Diagram 11.5

It is the function of 'the office' to collect, process, summarise and re-present in digestible format the information required by management in its decision-making role.

Top Tips

- Detailed guidance on farm business and financial planning can be found at www.gov.uk.
- Be consistent in the way the budget is prepared each year and keep budget notes and any useful background information to support calculations.
- Update management reports monthly.
- Monitor actual cash flow monthly and compare it to the budget.
- Look after the pennies and the pounds will look after themselves!

Chapter 12

Statutory and Assurance Records

Farming is bound by many rules and regulations that must be adhered to. These include legislative requirements, such as livestock movements; cross-compliance, a set of rules that must be followed in order to claim rural payments, such as Basic Payment Scheme (BPS) and Countryside Stewardship Schemes. There are various voluntary quality assurance schemes, which are coordinated under the Red Tractor Scheme, www.redtractor.org.uk and cover all farming sectors – beef, sheep, pigs, poultry, dairy, crops and produce.

Whether it is for rural payments or voluntary assurance, accurate and timely recordkeeping is required. Some rules are legally enforceable, and non-compliance can incur financial penalties or even court action. In order to comply with these, a farmer should have a recordkeeping system that is easily manageable and well-organised. Every farmer has a responsibility to ensure that his records are adequate and up-to-date and must be prepared to make the time to achieve this.

For the recording system to work, *entries need to be made on the day that the event happens.* A well-run and efficient recordkeeping system will cut down on duplication and undue stress incurred by a 'last-minute panic' prior to an inspection. There is no compulsory format in which the records should be presented but, whatever the system in use, an inspector from any inspecting body will expect the information required to be readily available. Assurance scheme providers usually offer guidance on recordkeeping together with the relevant record book/s.

Unless the administrator is working full-time on the farm, there are many records that are better dealt with by the farmer or enterprise manager. However, a working knowledge of the legislative requirements and recording systems is important, and an awareness of key deadlines and changes to rules is vital. As information relating to legislative requirements is constantly changing, it is advisable to use the internet for up-to-date information. There follows a brief overview of the vast amount of information available in relation to these key areas, more detailed information may be found at www.gov.uk.

Legislative Requirements and Recording Systems Fall into the Following Categories

- UK and EU legislation
- Rural payments
    - Basic Payment Scheme (BPS)
    - Countryside Stewardship schemes
- Farm Assurance
- Health and safety

UK and EU Legislation

Many of the recordkeeping requirements are underpinned by law and form the basis of the inspection framework. For example, the Disease Control (England) Order 2003 /Disease Control (Wales) Order 2003 sets out the Standing Movements Arrangements (SMA), which provide for the movement of cattle, sheep, goats, pigs and deer (with certain exceptions). EU regulations relating to cross-compliance may be found on www.gov.uk.

Similarly cattle are covered by several pieces of European Union legislation. The regulations include:

- Creating a system for cattle identification and registration through the use of double ear tagging, cattle passports, a fully operational computerised database and up-to-date farm registers.
- Developing a compulsory labelling system for all beef and beef products.
- Forming a minimum system of controls for spot inspections of British cattle holdings.

- Establishing time and reporting limits for cattle registration and ear tagging.

Rural Payments

The Basic Payment Scheme (BPS)

This is part of the Common Agricultural Policy (CAP) and is the European Union's main agricultural subsidy scheme. A farmer who meets the BPS rules can apply to the Rural Payments Agency (RPA)/Scottish Government (SGRPID)/Welsh Government (WG) for Basic Payments. These payments are not linked to production, giving flexibility in the running of the business. This principle is known as decoupling and the BPS is in the form of a single annual payment instead of individual support payments linked to arable crops and beef and sheep. However, in order to receive the BPS payment, there is a set of conditions that claimants have to meet and these are known as 'cross-compliance'.

Cross-compliance incorporates Statutory Management Requirements (SMRs) and Good Agricultural and Environmental Condition (GAEC). These cover: public, animal and plant health, environment, climate change, good agricultural condition of land and animal welfare. The SMR requirements reflect existing legal requirements in EU and UK law and GAECs are predominately ensuring the delivery of existing good practice across all farming activities. The current Guide to Cross-Compliance in England is only available online at www.gov.uk. It covers what must and must not be done in order to comply with the following.

Good Agricultural and Environmental Conditions (GAECs)

GAEC 1: Establishment of buffer strips along watercourses

GAEC 2: Water abstraction

GAEC 3: Groundwater

GAEC 4: Minimum soil cover

GAEC 5: Minimum land management reflecting site specific conditions to limit erosion

GAEC 6: Maintenance of soil organic matter level through appropriate practices, including a ban on burning arable stubble, except for plant health reasons

GAEC 7a: Boundaries
GAEC 7b: Public Rights of Way
GAEC 7c: Trees
GAEC 7d: Sites of Special Scientific Interest (SSSIs)
GAEC 7e: Ancient Monuments

Statutory Management Requirements (SMRs)

SMR 1: Reduce water pollution in Nitrate Vulnerable Zones (NVZs)
SMR 2: Wild birds
SMR 3: Habitats and species
SMR 4: Food and feed law
SMR 5: Restrictions on the use of substances having hormonal or thyrostatic action and beta-agonists in farm animals
SMR 6: Pig identification and registration
SMR 7: Cattle identification and registration
SMR 8: Sheep and goat identification
SMR 9: Prevention and control of transmissible spongiform encephalopathies (TSEs)
SMR 10: Plant Protection Products (PPPs)
SMR 11: Welfare of calves
SMR 12: Welfare of pigs
SMR 13: Animal welfare

Photo 12.1

Countryside Stewardship

Agri-environment schemes now fall under the umbrella of Countryside Stewardship, which is part of the Rural Development Programme for England (RDPE). Schemes are administered by Natural England and the Forestry Commission. Farmers and land managers may apply for funding, in addition to BPS, to deliver effective environmental management on their land. When existing schemes (ELS, HLS, etc.) finish, applications for a new scheme can be made under the following three main elements:

1. **Mid-tier:** Multi-year agreements for environmental improvements in the wider countryside. Existing schemes: Entry Level Stewardship (ELS), Organic Entry Level Stewardship (OELS) and Uplands Entry Level Stewardship (Uplands ELS) fall into this category.
2. **Higher tier:** Multi-year agreements for environmentally significant sites, commons and woodlands where more complex management requires delivery body support. Existing Higher Level Stewardship (HLS) schemes fall into this category.
3. **Capital grants:** A range of one- to two-year grants for hedgerows and boundaries, improving water quality, developing plans, woodland creation (establishment) and tree health.

Support

The Farming Advice Service (FAS) is funded by DEFRA to provide free, confidential advice to farmers and farming industry advisers to help them understand and meet requirements for cross-compliance 'greening', water protection and the sustainable use of pesticides. Contact details for FAS and an overview of the Rural Development Plan for England (RDPE) schemes can be found at www.gov.uk.

In Scotland, environment schemes can be accessed through the Agri-Environment Climate Scheme, see www.ruralpayments.org. For information about schemes relevant to Wales, go to www.gov.wales/rpwonline.

Inspections

European legislation requires inspection of a percentage of farm businesses submitting claims under the BPS and for other direct payments. Remote sensing may be used to check land areas; these are known as eligibility inspections, as opposed to inspections of cross-compliance requirements.

A farm can be selected for a number of inspections if payments are received. Inspections are carried out by England's RPA, Scotland's SGRPID and the Welsh Government's Rural Inspectorate of Wales (RIW). Other bodies carry out technical inspections and report to the relevant bodies. In the main, inspections are selected by risk assessment and would be carried out no more than once a year unless there were serious concerns. A business could also be selected for inspection by more than one inspecting body.

Advance notice of an inspection is given where possible, but it may be unannounced. The inspector will explain what they have found before leaving the farm. They will then write-up a full report to submit to RPA, who will use the report to assess if there has been any non-compliance with rules. Any breaches may result in a reduction to payment. The level of reduction is dependent upon the seriousness of the non-compliance.

Further details of the inspection process and penalty matrix can be found on the following websites:

England – www.gov.uk
Scotland – www.ruralpayments.org
Wales – www.gov.wales/rpwonline

Farm Assurance Schemes

Farm assurance schemes are voluntary schemes that establish production standards covering food safety, environmental protection, animal welfare, and health and safety issues and other characteristics deemed to be important by consumers. They include regular, independent checks on the producers that belong to the schemes, to ensure that they follow the rules. They are designed to assure consumers that farmers and growers are producing food to meet all legal standards and an agreed set of standards of good agricultural practice (see www.redtractor.org.uk).

Physical Inspections

Inspectors make routine farm visits at least every 18 months. These routine visits are validated by a series of random spot-checks aimed at ensuring consistent application of the standards. No producer can sell products as 'Farm Assured' until an independent inspector has visited the farm to check he meets the standards. Details of the standards are on the Red Tractor Scheme website at www.redtractor.org.uk.

For example, the standards for a livestock farm cover the following areas and inspectors will inspect against these standards:

- Identification and traceability
- Farm animal management
- Environment and hygiene management
- Feed composition, storage and use
- Housing and handling facilities
- Medicines and veterinary treatments
- Transport of livestock
- Expansion of home-mixing standards for feed
- Herd/flock health plans
- Waste management plans.

There are additional assurance schemes being operated particularly for produce. These are usually implemented by the larger supermarkets and will be a requirement of obtaining a sales contract with them.

However, if a farmer is not a member of an assurance scheme there is still basic legislation that ensures that minimum standards are adhered to, and he will still be liable to be inspected by the regulatory bodies.

Health and Safety

This legislation is far-reaching and the structure and scope of a business will determine what legislation is applicable. Guidance and support is available and a service provider specialising in agriculture is recommended. For do-it-yourself situations, a comprehensive guide called 'Farmwise' is available from www.hse.gov.uk.

An up-to-date Health and Safety Law poster together with a current employer's liability insurance certificate must be displayed in an appropriate place where staff will see it. If there are five or more employees, there must also be a written policy on health and safety. More information is available from www.gov.uk.

A well-run business will have in place practical measures, demonstrating that they are concerned with staff welfare, such as providing them with an emergency plan and contact list. Laminated copies of this important information should be supplied to all staff, and may also be placed in farm vehicles, unit offices and restrooms and should include:

- Emergency telephone numbers, first aider, doctor, nearest hospital, etc.
- Information for emergency services – clear directions to the farm with OS grid reference numbers.
- Location of isolation points: gas and electricity and nearest alternative water supply.
- Location of first aid kits and fire extinguishers.

A range of useful templates including emergency plan and contact list can be found at www.redtractor.org.uk.

Photo 12.2

Recordkeeping

Practicalities of Getting Information from Farm to Office

Alongside the office systems, there needs to be a system for collecting the information from day-to-day farm activities that are required to be recorded for legislative, cross-compliance and assurance scheme purposes. The most basic, but probably still the most effective, means of collecting information is ensuring that family members, staff and contractors are equipped with a notebook/ small diary/cab card. For example, a diary/cab card could be put into each tractor/self-propelled machine. By having a diary/cab card in each machine, the driver can record the required information – such as cultivations and weather conditions – on a field-by-field basis.

At the grain store, a diary can be used to record all loads going out. These entries must correspond with the collection notes left by the lorry, which then check with the purchasers' self-billing invoices. Information from the diary/cab card can then be collected in the farm office and used to maintain the necessary paper records or, increasingly now, onto computer programs. From the entries, it is possible to update the program, which will then produce reports set out in the form that would satisfy any inspector.

Pen and paper is still a tried and tested method of recording, but the advance in technology and the move to more efficient methods of data capture and reporting have led to the increased use of computerised systems. This continues to be a fast developing area. For example:

Crop records: Technological developments have made light work of field records. Crop protection or fertiliser recommendations supplied by an agronomist electronically may be implemented by the tractor driver/ spray operator and field records held in a computerised crop-recording program can be updated. The addition of the application date and weather conditions completes the record and removes the need to enter data manually from handwritten instructions.

Livestock records: With the introduction of Electronic Identification (EID), it is now easy to record information required by legislation (e.g., movements) and useful management information on individual animals. EID is based on electronic devices attached to an animal and readers used by the herdsman/shepherd. The device used is a microchip

Photo 12.3

in either an ear-tag or a ceramic bolus. The bolus is swallowed by the animal (only used with ruminants) and remains in the animal stomach for its lifetime. The reader in a farm situation may be a handheld device called a stick-reader; otherwise hands-free devices are available for attachment to weigh crates and other stock handling equipment.

Data transfer: Stick-readers are principally used for reading EIDs, which can be transmitted (via a wireless link) to another device such as a mobile phone. Cattle birth and movement records may then be sent directly to the Cattle Tracing System (CTS), see www.bcms.gov.uk. The same device can be used for reading and transmitting other individual animal records to a livestock recording program, such as births, veterinary treatments and weights, again removing the need for manual records.

Farm/Business Identification

In order to be eligible for any government support payment, the holding must be registered correctly with the relevant farming bodies in England, Scotland

and Wales. For many years a 'holding number' was the only unique administrative reference number allocated to a farming unit. However, today, a farm may have a range of reference numbers allocated by the following bodies.

- England – DEFRA and Rural Payment Agency (RPA): www.gov.uk.
- Scotland – Scottish Government Rural Payments and Services (SGRPS): www.ruralpayments.org.
- Wales – Welsh Government (WG): www.gov.wales/rpwonline.

Agricultural holdings are identified by:

- **Holding number:** A unique number for the holding. Referred to as a CPH number – stands for county/parish/holding and will consist of 2/3/4 digits, i.e., 17/107/0015.
- **Single business identifier (SBI):** The SBI is for the business. In Scotland, this is called a business reference number (BRN) and in Wales, it is a customer reference number (CRN).
- **Personal identifier (PI):** This is a personal number. For example, where more than one person runs the farm, then each will have a PI number needed when contacting the Rural Payments Agency (RPA). Farm manager/administrator may also have a PI number if they are required to talk to the RPA. Currently PI Numbers only apply in England and Wales.
- **Vendor number:** This is a unique trader registration number allocated by the RPA in order to receive payments.
- **Customer reference number (CRN):** This will be needed together with a password to logon to the online BPS claim system (Basic Payment Scheme).
- **Herd/flock number:** This number identifies the livestock and directly relates to an individual holding. Pigs have a different herd number format to sheep, goats and cattle.
- **Rural Land Registry (RLR):** This is part of the RPA and is responsible for maintaining digital maps showing the location and land areas of each holding. To be eligible for payments through the Basic Payment Scheme (BPS) or the other environmental schemes, all land must be registered with RLR now known as Land Management System (LMR).

It is useful to have an easily accessible record of all frequently used reference numbers and the farm size.

Farm Holding No (CPH)	17/301/0123
Single Business Identifier (SBI)	11223344556
Personal Identifier (PI)	106333444
Vendor Number	356654
Herd Number	UK312345
Flock Number	UK312345
Farm Assurance Scheme Number (Livestock)	05566
Farm Assurance Scheme Number (Crops)	03344
Farm Size Acres	300 acres
Farm Size Hectares	121.4 hectares

Diagram 12.1

Recordkeeping Requirements

As each devolved country can have its own particular requirements, it important to be aware of these and understand the variations in the rules, etc. This chapter deals mainly with the English recordkeeping requirements with links to the appropriate bodies in Scotland and Wales. The guide to cross-compliance in England is only available. It is possible to drill down into each SMR online to find current recordkeeping requirements in detail.

For simplicity, records are subdivided into three main areas:

1. Livestock
2. Crop
3. Generic (health and safety, etc.)

Livestock Records – SMR4, SMR5, SMR6, SMR7, SMR8, SMR9

For information pertaining to Scotland see www.ruralpayments.org. For Wales, go to www.gov.wales/rpwonline.

The basic rules for all species are:

Flock/herd number: By law, all land where cattle/sheep/goats/pigs are kept must be registered and a flock herd number must be obtained from the Animal and Plant Health Agency (APHA).
Cattle keepers: A herd number should be obtained from the Animal and Plant Health Agency (APHA), and keepers should be registered with the British Cattle Movement Service (BCMS, www.bcms.gov.uk). The Rural Payments Agency (RPA) through the British Cattle Movement Service (BCMS) runs Great Britain's Cattle Tracing System (CTS) database.
Pigs, sheep and goats: These should be registered with the following organisations depending on country:

- England – keepers should register with www.defra.gov.uk – Animal and Plant Health Agency (APHA) office, which will allocate a flock/herd number (even if only one animal is kept).
- Scotland – keepers should be registered with SGRPID, www.gov.scot. The local Animal Health Divisional Office (AHDO) will allocate a flock/herd number.
- Wales – keepers should be registered with the local Welsh Government Rural Affairs Office. See www.gov.wales/rpwonline.

Movement Records and Licences

Farmers typically move livestock under a general licence (excluding poultry). All movements (except where new ten-mile rule applies, see below) are recorded, which means that should there be a disease outbreak, movements can be traced quickly. Records can be manual or electronic. Cattle and pigs now also have central data recording systems.

New Ten-Mile Rule – Changes Being Phased from July 2016 to Autumn 2017

These changes will simplify livestock movement rules by removing a number of complex exemptions and by making the rules the same for all species. For many keepers, the changes to the ten-mile rule will reduce the work associated with movement reporting and standstills and give a better understanding of where livestock are.

Under the new system, farmers will be able to move their animals around their land, within a ten-mile radius, without the need to report and without standstills (see below). Farmers affected by these changes will be contacted and may apply to register all the land used, within a ten-mile radius (place of business), under the same CPH (County Parish Holding) number. This will cover land used permanently or on a temporary basis. Full details can be found at www.gov.uk.

Standstill Restrictions (Subject to Revision under the New Ten-Mile Rule)

To prevent the spread of disease, there are a number of 'standstills' that apply to animal movements. Legislation dictates that movements of cattle, sheep, goats and pigs off a holding can only take place after complying with a standstill period. The standstill period where *cattle*, *sheep and goats* are moved onto a holding is *six days* (excluding the day of arrival – i.e., Saturday on equals Saturday off).

Where *pigs* are moved onto a holding the standstill period is *20 days* (excluding the day of arrival). If cattle, sheep or goats are moved onto a holding then no cattle, sheep, goats or pigs can be moved off the holding within six days of that movement. For example, cattle, sheep or goats brought onto the holding on a Saturday will mean that cattle, sheep, goats or pigs will be unable to move off the holding until the following Saturday. If pigs move onto a holding, then no pigs can be moved off the holding within 20 days. However, any cattle, sheep or goats on that holding will be able to move after six days of the movement. Where pigs move on to a holding on the 1st of the month, cattle sheep and goats can move off on the 8th and pigs can move off on the 22nd of the month.

Exemptions and Variations

Exemptions are allowed from the standstill periods if livestock are moved direct to a slaughterhouse or a dedicated slaughter market. Movement restrictions are slightly different in Scotland. For cattle, sheep and goats the standstill period is 13 days and for pigs it is the same as England, i.e., 20 days. There are also special rules relating to movements to and from agricultural shows. If in doubt, check with the Animal Health/Trading Standards office managed by the local authority, which deals with animal movements. More information can be found on the DEFRA website.

Medicine Records

These are compulsory for stock that might enter the food chain. A record must be kept of drugs purchased, used, batch number, date of commencement and completion of treatment, who administered the drug and, most importantly, the 'withdrawal date'. This is the date x days after completion of the treatment, where a date is specified for each individual medicine and treatment regime; it represents the stage at which the residues in meat or milk will have fallen to or below the maximum residue level (MRL) for it to be acceptable for human consumption. Only authorised animal medicines should be used. Records must be kept for at least five years following the administration or other disposal of the product.

Fallen stock

- A farmer is not allowed to bury fallen stock unless there is a derogation (e.g., if farming on the Isle of Wight).
- Fallen stock should be disposed of to a licensed operator.
- Fallen stock scheme – producers pay to join the scheme and then pay a reduced rate per animal disposed of.
- Records must be completed when transporting animal by-products (livestock carcasses/parts of carcasses that may have died on the farm premises).
- A receipt for the fallen stock must be produced in triplicate; the receiver retains the original, while the producer and the transporter also keep a copy available for inspection.
- Records must be retained for two years.

An *Animal Transport Certificate* must accompany all animals in transit, journeys fall into the following categories:

- **Short journeys (Type 1) – up to 65km (40 miles):** Anyone transporting animals must comply with rules on the *General Conditions for the Transport of Animals*, covering fitness to travel, means of transport, transport practices etc. Farmers transporting their own animals, in their own vehicles less than 50km need only comply with the *General Conditions for the Transport of Animals*.
- **Long journeys (Type 2) – for over 65km and up to eight hours duration:** As above plus:

  - Must hold a valid Transporter Authorisation.
  - Drivers and attendant responsible for the transport of cattle, sheep, pigs, goats, horses or poultry must hold a valid *Certificate of Competence*.
- **Transporting animals over eight hours:** As above plus:
  - Road vehicles and containers subject to inspection.
  - A journey log must accompany animals being transported to other EU Member States.

Records must be retained for six months. Full guidance notes may be downloaded from www.gov.uk.

More Rules Specific to Each Species

Cattle Records SMR7

A record must be kept of all cattle births, movements and deaths, which include details of:

- Ear-tag number
- Date of birth
- Sex
- Breed
- Identity of dam
- Date of movements 'on' and 'off' your holding
- Details of where the animal has moved to or from
- Date of death.

A record must be kept of all cattle movements, even if they do not need to be reported to BCMS. Keepers can apply to BCMS to have a link (of holdings) in CTS.

Deadlines

Information must be entered into the records within the following deadlines:

- Thirty-six hours in the case of movements 'on' or 'off' a holding.
- Seven days for the birth of a dairy animal.
- Thirty days for the birth of all other cattle.
- Seven days for a death.
- Thirty-six hours for replacement ear tags.

Notification to BCMS has the following deadlines:

- Cattle movements – within three days of the movement taking place.
- Death of bovine animal – return passport within seven days.
- Application for passport – within a total of 27 days from birth.

It is important that records are updated as soon as possible after the event to ensure that they are accurate.

Record Retention

Cattle records must be kept for ten years from the end of the calendar year in which the last entry was made. The records may be paper or stored on a computer.

Sheep & Goats, SMR 8

Records

A holding register must be kept and the following should be recorded within it:

- Holding details
- Tag replacements
- All sheep and goat movements on and off your holding
- Date of identification and deaths
- Annual count of animals on the holding as at 1 December each year (Scotland and Wales, 1 January)
- Individual records of sheep and goats born or identified after 31 December 2009
- A separate register must be kept for each individual holding.

Deadlines

- All movements of sheep and goats must be reported to the Animal Reporting and Movement Service (ARAMS) within three days of the movement. Movements are reported on movement document – ARAMS 1.
- All movements must be recorded in the Holding Register within 36 hours (48 hours Scotland) of the movement taking place.

Record Retention

- A copy of the ARAMS 1 must be retained for three years from the date of movement onto the holding.
- The Holding Register must be retained for three years from the date of the last entry.

Scotland

The Scottish Animal Movement Unit (SAMU) See www.gov.scot which operates the sheep, goat and pig movements database in Scotland and all movements of sheep and goats must be notified to SAMU.

Wool Board Registration

If more than four sheep are kept, the business should be registered with the British Wool Marketing Board in order to sell wool through them (www.britishwool.org.uk).

Safe Use and Disposal of Sheep Dip

Purchase and use of sheep dip requires the purchaser to hold a valid certificate of competence in the safe use of sheep dip. Details of who holds this should be kept in a training record file.

Use of sheep dip should be detailed in the veterinary medicine records, including the date, product, supplier, batch number of product, batch of animals treated and withdrawal period.

It is also wise to record the time and place where sheep are treated. This will provide useful evidence in the event of a pollution incident being investigated and will act as a reminder to observe any withdrawal periods for sheep going to slaughter.

Regarding the disposal of sheep dip, it is a legal requirement under the Environmental Permitting Regulations (EPRs) to hold a permit for the disposal of waste sheep dip/pesticide on to land. A condition of the permit is that records must be kept of the types of substances disposed of, volumes, dates and location of disposal.

Pigs SMR 6

- When a holding is registered for pig-keeping, a herd mark will be automatically created.

- Once a year, a record of the maximum number of pigs normally kept on the holding should be made.
- All movements of pigs must be pre-notified electronically through the eAML2 (www.eaml2.org.uk) system or by telephoning the Meat and Livestock Commercial Services Limited (MLCSL).
- An on-farm movement record should also be kept in format shown in Diagram 12.2.

Date of movement	Identification mark, slapmark or temporary mark	Number of pigs	Holding from which moved	Holding to which moved
01/05/xx	*Slapmark on both shoulders AB1234*	*5*	*My holding Full address CPH Number*	*Mr New holding Full address CPH Number*

Diagram 12.2

Pig enterprises with places for 750 sows or 2,000 plus growing pigs require a permit under Environmental Permitting Regulations (EPR). There are a number of recordkeeping requirements under EPR – for further details see www.gov.uk.

Deadlines

All movements of pigs must be recorded within 36 hours of the movement. The recipient (new keeper) must notify MLCSL electronically within three days of receiving the pigs or by phone, fax or in writing.

Record Retention

All records must be kept for six years even if pigs are no longer kept and must be available for inspection on demand.

Scotland – for record requirements go to www.gov.scot.

Poultry

If more than 50 birds are kept, the flock must be registered with the GB Poultry Register. The register must be kept up-to-date to help Animal and Plant Health Agency (APHA) manage potential disease outbreak. It is recommended to register for disease alerts. Poultry enterprises with places for

40,000 birds and over require a permit under Environmental Permitting Regulations (EPR). There are a number of recordkeeping requirements under EPR – for further details see www.gov.uk. Any person who is engaged in the transport or marketing of poultry or eggs, including slaughterers and auctioneers, must also keep records.

Records

The following needs to be recorded where 250 or more birds are kept:

- Date and place poultry were obtained
- Species and description
- Name and address of the person from whom they were obtained
- Date they left the premises
- Destination on leaving the premises (if known)
- Purpose for which they left the premises
- Name and address of the person to whom they were transferred.

Record Retention

The records, once made, must be kept for at least 12 months and be produced to an inspector.

Strict legislation and codes of practice on the catching and transport of poultry must be adhered to and more information can be found at www.britishpoultry.org.uk and www.gov.uk.

Crop Records SMR 1 and SMR 10

As with the livestock sector, arable farmers must also follow cross-compliance rules in order to receive support under the Basic Payment Scheme. The majority of farms are signed up to the relevant crop and produce assurance schemes and comply with their standards. Many of these rules and standards come with very clear environmental and food safety objectives and outcomes. Explanations of some of the drivers for these are given later in this chapter.

Recordkeeping

The keeping of accurate and detailed records is an essential part of arable, grassland, fruit and vegetable production and is a basic necessity of good management. As with livestock production, there are recordkeeping requirements under legislation, farm payment schemes and quality assurance schemes.

Again, this is an area of recordkeeping that has benefitted hugely from the introduction of computer programs and with the introduction of Global Positioning Systems (GPS), precision farming and sophisticated computer programs in tractor cabs. There are also basic templates and software packages for recordkeeping provided by assurance schemes and government departments.

Field Records

The basic requirement is to keep a record of all the activities associated with the crop from autumn cultivations and sowing through to harvest. Useful templates for crop recording can be downloaded from www.redtractor.org.uk.

Regarding *pesticide spray records* (which includes herbicide, fungicide, insecticide, growth regulators and granular applications), HSE is the regulatory authority in the UK for pesticide products. Pesticides can be split into two categories, those used in agriculture, horticultural and home garden, and those used for public hygiene (www.hse.gov.uk).

Spray recordkeeping need not involve vast amounts of paperwork, but does require some organisation and management. Sound systems that aid the capture of the basic data are vital and then in turn this information can be easily transferred to computer programs for management purposes as well as achieving the necessary legal requirements.

Regarding *fertiliser records* (including organic manure applications and treated sewage sludge products), keeping of accurate records of fertiliser usage helps farmers to apply best practice when growing crops and minimises the unnecessary and wasteful applications of nutrients.

Records relating to the planned and applied use of nitrogen fertiliser and organic manures are a requirement for farms in a Nitrate Vulnerable Zone (NVZ) area under the Nitrate Pollution Prevention and are included in Cross-Compliance SMR1 to reduce the pollution of waters caused or induced by nitrates from agricultural sources and to prevent further pollution. The fertiliser application record is a key record required to be kept for the Red Tractor Farm Assurance Fresh Produce and Assured Crop Schemes. For more details see NVZ records.

Summary of Records to be kept		
Type of record	**Time kept for**	**Reason for records**
Pesticide treatments	Refer to hse.gov.uk	To show that pesticides have been used appropriately, safely and legally. By law, food and feed producers must keep these records. To help with good management. To provide other people with important information, especially in emergencies when people, animals or the environment have accidentally been contaminated, or crops have been damaged. To help keep track of periods when crops cannot be harvested and people or animals cannot enter a treated area. To meet the specific conditions of crop assurance schemes or the woodland assurance standard.
Pesticide store Records	Until updated	To give an accurate list of the contents of the chemical store in an emergency and to help with stock control.
LERAPs (Local Environment Risk Assessment for Pesticides)	Three years	To show that the conditions of the LERAP and agri-environment or stewardship schemes have been met. To allow for the assessment of the effectiveness of a particular pesticide. To meet the specific conditions of crop assurance schemes or the woodland assurance standard.
COSHH (Control of Substances Hazardous to Health) **Assessment and Environmental Risk Assessment**	Until revised	To show that all risks to people and the environment have been adequately assessed. To provide evidence that the legal obligations to protect people and the environment have been met. To confirm that the appropriate certificates or permission have been obtained (for example, authorisations under the Groundwater Regulations).
Maintenance, inspection and testing of measures to control exposure	Five years	To confirm that measures to control exposure are working effectively. To show that employers have met their legal duty to maintain, inspect and test engineering controls and respiratory protective equipment.
Monitoring exposure in the workplace (general samples from the workplace)	Five years	To confirm that the level of exposure at work is acceptable. To show that employers have achieved and maintained adequate control of exposure to dangerous substances.
Monitoring exposure of individual, identifiable people	Forty years	To confirm that the level of exposure at work is acceptable. To show that employers have achieved and maintained adequate control of exposure to dangerous substances.
Health surveillance	Forty years	To identify any negative health effects resulting from exposure to dangerous substances at work and to show that employers have met any legal conditions to carry out health surveillance of their employees.
Disposal Records	Two or three years	To show that waste has been handled and disposed of safely and legally.

Diagram 12.3

Example of a Fertiliser Applications Record					
Field name/Number/ID:	Ash Tree Field	**Area (Ha):**	8.20	**Crop Year:**	20xx
Soil Type:	Medium Clay	**Current Crop:**	Potatoes	**Previous Crop:**	Wheat
Variety:	Russet	**Seed dressing:**		**Purchased seed lot no:**	SV 6328
Sowing Date:	15/3/20xx	**Comments**		**Home saved seed:**	**~~Yes~~ / No**
Date of last soil analysis (Retain copies for inspection)	September 20xx			**Harvest Date**	
Previous crop residues	**N index:** **PH:**		**P index:**		**K index**
Crop Needs – as per RB 209	160		170		300

Fertiliser and organic manure applications including treated sewage sludge products

Date	Product	Rate	N	P	K	S	Notes	Operator
1/3/20/xx	FYM	40T/ha	48	76	288			

Diagram 12.4

Fertiliser Storage Records

Manufactured fertiliser is a valuable product for farmers and growers but is potentially dangerous in the wrong hands. The storage and security of fertiliser is therefore of paramount importance. Alongside the storage rules are a number of recordkeeping requirements:

- Record fertiliser deliveries and usage.
- Carry out regular stock checks and report immediately any stock discrepancy or loss to the police.
- Record any manufacturer code numbers from the bags and, if available, the number of the detonation resistance certificate.

Further information is available in the HSE booklet 'Storing and Handling Ammonium Nitrate' (www.hse.gov.uk).

Water Abstraction

A farm business that is abstracting more than 20 cubic metres per day of water requires an abstraction licence from the Environment Agency and needs to keep a record of the meter-reading of individual meters showing meter number and water used and submit to EA annually.

Safe Disposal of Waste Pesticide

Disposal of pesticide washings is a legal requirement under the Environmental Permitting Regulations (EPR) to hold a permit for the disposal of waste pesticide on to land. A condition of the permit is that records must be kept of the types of substances disposed of, volumes, dates and location of disposal.

Generic Records (Livestock and Crop Production)

There are a number of records that are required to be kept whether a farm is producing livestock, corn, fruit, vegetables or has mixed enterprises; these must be kept up-to-date, available for inspection and include:

- A visitor's book.
- Machinery service records – for example, fertiliser sprayer calibration.
- Training records – list of literature that has to be kept – evidenced at inspections.

Nitrate-Vulnerable Zone Records SMR 1

For those farming in an NVZ designated area, it is important that the principles are understood and the records are maintained. This is a key area to be checked if the farm is subject to a cross-compliance inspection. Maps denoting the designated areas can be found at:

- England – www.defra.gov.uk.
- Scotland – www.gov.scot.
- Wales – www.gov.wales/rpwonline.

If a farm is in an NVZ or has land in a designated area, the risk of water pollution must be assessed. A Nutrient Management Plan will help do this and meet NVZ regulations, which cover:

- The use of livestock/organic manure and manufactured nitrogen fertilisers to ensure that no more nitrogen is applied to the crops than is required. The N max limit for available nitrogen (in livestock manures and fertilisers) applied to certain crop types (as listed in the regulations) must not be breached.
- A risk map must be produced for any land where organic manure is intended to be spread and stored.
- Closed periods and spreading controls for spreading manufactured nitrogen fertilisers and organic manures must be adhered to.
- Nitrogen produced by livestock must be calculated and recorded each year along with the hectarage of farm to ensure that the livestock manure N (nitrogen) farm limit is not exceeded.
- The farm must provide adequate storage capacity for slurry produced from livestock. A record of the numbers of livestock kept on the farm during the slurry storage period must be kept.
- Keep records of the nitrogen applied to each of the fields.
- Records must be kept for at least five years.

The 'Tried & Tested Nutrient Management Plan' booklet is an aid to nutrient planning and recording. It can be downloaded from www.nutrientmanagement.org. Record sheets are also available from this site in various formats including an Excel workbook with useful pre-set calculators. There are extra rules for storing organic manures; the following guides may be found on www.gov.uk – Nitrate Vulnerable Zones and Storing Silage, Slurry and Agricultural Fuel Oil.

Vermin/Rodent Control

Control of vermin (including birds, rodents and insects) and other animals (including cats and dogs) is vital to prevent contamination of animal feed or harvested crops. This is important in the production of safe food, managing biosecurity and preventing the spread of disease. There is a requirement of some of the farm assurance schemes to maintain a record to evidence regular inspection and control of any vermin present.

Disposal of Agricultural Waste

Agricultural waste is any substance or object from premises used for agriculture or horticulture, which the holder discards, intends to discard or is required to discard. It is waste specifically generated by agricultural activities.

Some examples of agricultural waste are:

- Empty pesticide containers
- Old silage wrap
- Out of date medicines and wormers
- Used tyres
- Surplus milk.

What Needs to be Done

Find out about registering exemptions, moving waste, rules and regulations relating to manures and slurries, and dealing with hazardous waste on the farm.

Registering Exemptions

Usually, any waste treatment, recovery or disposal activity needs to be carried out under an **Environmental Permit**. However, some low-risk, small-scale and less polluting activities may be exempt from permitting.

These activities are known as **'exemptions'** and need to be registered online with the Environment Agency. Most farmers are carrying out activities that require exemptions to be registered. For example:

- Burning plant material.
- Using old tyres on silage clamps.
- Spreading waste milk on land.

There are a number of other farming activities involving the management and disposal of waste that have to comply with rules and regulations (as listed below):

- Spreading waste on farmland.
- Moving waste.
- Dealing with hazardous waste.

Further information can be found on the following websites:

- England/Wales – www.gov.uk.
- Scotland – www.sepa.org.uk.

Background Information
It is obvious that there is a requirement for farming to comply with rules and regulations. In the livestock sector, much of this has been imposed since the BSE crisis and foot-and-mouth outbreaks, in order to raise the standard of traceability and food safety.

Alongside this is the environmental agenda, which covers a wide variety of areas including the maintenance and conservation of habitats, landscape and wildlife features and the effects of diffuse pollution. In an effort to improve these areas, farmers are required to work with a number of initiatives, rules (some of which are which are enforced by law).

Top Tips

- Keep hard copies on file of all documentation submitted on paper or online and make file notes of all telephone conversations with government departments and assurance scheme providers.
- Use a diary/desk-top tasks manager for deadline reminders.
- Keep up-to-date with changes in cross-compliance rules and assurance scheme standards, make sure that farmers (employers and clients) are aware of updates as well.
- Have a workable system that ensures that day-to-day information gets into the office promptly.
- The internet is the prime information source point – get familiar with the relevant websites, use them regularly and sign up for email alerts.

Who Does What?

Through the course of this chapter, a number of organisations have been referred to as they are involved with the setting up, monitoring and inspections of farming regulation – the key ones are listed in Figure 12.6 on page 157.

Water Framework Directive	European legislation that provides the framework for the water bodies across Europe
Campaign for the Farmed Environment	The aim is to exceed the environmental benefits offered by set-aside by establishing a coordinated and engaging "Campaign for the Farmed Environment". The Campaign encourages farmers to manage their land well for the environment on a voluntary basis – for example, by taking land out of production, putting in place buffer strips or managing more risky crops to prevent erosion and run off.
Groundwater Regulations	Regulations that apply to the groundwater that is below the surface of the ground – below the water table and in direct contact with the ground & subsoil. Rules to be aware of are those that apply to the disposal of sheep dip and pesticides.
Waste Regulations	The regulations affect businesses that produce waste, import or export waste, carry or transport waste, keep or store waste, treat waste, dispose of waste, operate as waste brokers or dealers.
Food & Environment Protection Act	Statutory powers to control pesticides with regard to:- Protecting the health of human beings, creatures and plants; safeguarding the environment; secure safe, efficient and humane methods of controlling pests; making information about pesticides available to the public.
Voluntary Initiative	The farming/crop production industry working with the Government has developed an initiative to achieve the environmental benefits as an alternative to a pesticide tax.
National Register for Spray Operators	Central register of spray operators who agree to participate in a Continual Professional Development Programme (CPD).
National Sprayer Testing Scheme	Scheme which ensures efficiency of sprayers & gives a certificate as evidence of machine soundness.
Crop Protection Management Plan	The production of a plan by a farmer that considers the environmental impact of his activities and the steps that are taken to reduce it.
Catchment Sensitive Farming	An initiative to reduce the agricultural sources of diffuse pollution within designated water catchments.
Local Environment Risk Assessment for Pesticides	For certain pesticides it is a requirement to leave 'buffer zones' (untreated areas) to protect water and anything living in it when there is an application of pesticide with a ground water or broadcast air assisted sprayer.
Nitrate Vulnerable Zone (NVZ's)	Areas in England, Scotland & Wales that have been designated to protect all the surface & groundwater against nitrate pollution and lower those areas that are high in nitrates.

Diagram 12.5 Rules and regulations

Department for Environment Food & Rural Affairs (Defra)	The Government Department that provides the support for British Agriculture dealing with the whole food chain and ensures that there is a secure, healthy and environmentally sustainable supply of food with improved standards of animal welfare.
Rural Payment Agency (RPA)	This organisation is an Executive Agency of Defra and provides services such as rural payments, rural inspections and livestock tracing.
Scottish Government Rural Payments and Inspections Directorate (SGRPID)	Scottish Government Department which supports Scottish Agriculture as Defra does in England and it also combines rural payments, rural inspections and livestock tracing.
Welsh Government	The Welsh Government supports the countryside and rural communities, encouraging the sustainable management of agriculture and the environment. It also combines rural payments, rural inspections and livestock tracing.
Natural England (NE)	The government's adviser on the natural environment which provides practical advice, grounded in science, on how best to safeguard England's natural wealth for the benefit of everyone. The organisation works with farmers and land managers; business and industry; planners and developers; national and local government; interest groups and local communities to help them improve their local environment.
Environment Agency (EA)	Regulates farming through regimes such as Nitrate Vulnerable Zone (NVZ) compliance, water abstraction licences, discharge consents and farm waste regulations.
Animal & Plant Health Agency (APHA)	The home of Animal and Plant Health Agency on GOV.UK. Works to safeguard animal and plant health for the benefit of people, the environment and the economy.
Local Authorities	Where farmers report livestock movements to via the Animal Movement Licence system and also inspect and regulate the Standard Movement Arrangements.
Veterinary Medicine Directorate (VMD)	Local Authorities are responsible for the Animal Movement Licence System. They also inspect and regulate the Standard Movement Arrangements.
The Red Tractor Assurance Schemes	The organisation that sets out to maintain, develop and promote integrated assurance standards for the benefit of its assured members within the whole of the food industry. It sets standards for farms as well as other critical links in the food supply chain both pre and post farm.

Diagram 12.6 Key organisations

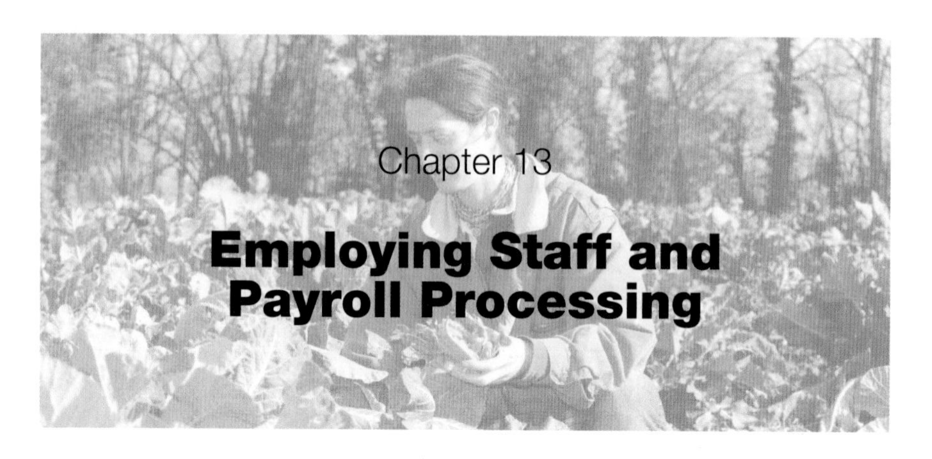

Chapter 13

Employing Staff and Payroll Processing

Many farm businesses rely on the labour of family members and hire in contractors when extra help is needed, whereas a large-scale vegetable producer may employ a varying number of 'field' staff throughout the year and a full-time administrator to cover payroll processing and associated staff records. In this chapter, we aim to give an overview of the administrative responsibilities of a small employer of agricultural workers.

Preparing to Take On a New Employee

Writing a job description is a useful starting point and is also helpful when it comes to advertising the position. It is acceptable to specify the necessary qualifications and experience in the advert but be careful not to include any discriminatory terms. In planning interviews, which should ideally involve the person in charge and other colleagues, a range of questions should be prepared beforehand. References from previous employers are strongly recommended. Recruiting unsuitable employees can be very costly to a business. Employee fraud statistics indicate that one in four businesses suffer fraud and small and medium-sized enterprises (SMEs) are the most common targets. Advisory, Conciliation and Arbitration Service (ACAS) have a range of useful templates, forms and checklists to help with the process of hiring staff (www.acas.org.uk).

Agricultural Wages Board, AWB – (England) Abolished 30 September 2013

The abolition of the Agricultural Wages Board (AWB) in England and subsequent revocation of the Agricultural Wages (England and Wales) Order 2012 on 30 September 2013 puts agricultural and horticultural employees in England on the same footing as employees in all other industries and comprehensive information on all the topics covered in this chapter can be found at www.gov.uk. Agricultural workers taken on before the AWB was abolished continue to be employed under their original terms and conditions, unless there is mutual agreement between the employee and the employer to vary these terms. However, if they move to a new employer, they must be paid at least the National Minimum Wage (NMW) or National Living Wage (NLW).

Scotland and Wales AWBs Were Not Abolished

Search on the relevant government websites for up-to-date AWBs, setting out minimum rates of pay, holiday entitlement and sick pay, etc.:

- Agricultural Wages (Scotland) Order and Guide – www.gov.scot.
- Agricultural Wages (Wales) Order – www.gov.wales.

Rates of Pay: Employment Started on or After 1 October 2013

The National Minimum Wage (NMW) is the minimum pay per hour that almost all workers are entitled to by law from April 2016. Rates fall into the following six categories and are revised annually in October. Accommodation provided by an employer can be taken into account when calculating NMW, the rates can also be found at www.gov.uk and are also revised in October.

- Aged 25 and above (National Living Wage, NLW).
- Aged 21 to 24 inclusive.
- Aged 18 to 20 inclusive.
- Aged under 18 (but above compulsory school leaving age).
- Apprentices aged under 19.

- Apprentices aged 19 and over, but in the first year of their apprenticeship.

Employment Started Before 1 October 2013

Workers, including those supplied by a gangmaster, are still entitled to the terms and conditions of their contract. For example, this might mean workers are entitled to overtime rates, agricultural sick pay and dog allowance. Where accommodation is provided in a contract, workers can continue living in that accommodation. These entitlements and any other terms and conditions already agreed will continue to apply unless the contract is changed by mutual agreement or it finishes.

Workers must always be paid at least the appropriate NMW. An agricultural worker's rate must be raised if it ever falls below the minimum.

Holiday Entitlement

Holiday entitlement for a farm worker employed before the abolition of the AWB (30 September 2013), working a standard 39-hour week is entitled to 31 days (including some Bank Holidays). Workers taken on from 1 October 2013 are entitled to at least the national minimum, which is currently 5.6 weeks per year (including some Bank Holidays). The holiday year runs from 1 October. There are holiday entitlement calculators at www.gov.uk for both of these categories. The calculators also deal with the overtime factor, see below. In Scotland, holiday entitlement and rates are set by the SAWB, with the holiday year running from 1 January. ACAS also publishes a comprehensive guide, 'Holidays and Holiday Pay', at www.acas.org.uk.

Holiday Pay and the Consideration of Overtime

Where a worker's pay does not vary – for example, they are salaried or never work overtime – it is simply a matter of paying the normal rate of pay for days taken as holiday. If a Bank Holiday is worked (say, eight hours) the worker is paid eight hours at the overtime rate and has a day off in lieu. Recent case law means that guaranteed and non-guaranteed overtime should be taken into account when calculating holiday pay. This is done by working out the average weekly gross pay over the preceding 12 weeks prior to the start of the holiday.

- **Guaranteed overtime** is where the employment contract requires both the employer to offer the overtime and the worker to work it.
- **Non-guaranteed overtime** is where the employer is not required to offer it but the worker is required to work it once offered.
- **Voluntary overtime** is work the employer may ask the worker to do and it is the worker's choice whether to work it.

This particular area of employment law is likely to change, so it is important to keep up-to-date rather than run the risk of facing claims from workers, which can be backdated.

Casual Workers with Irregular Hours

For these workers, it is often easiest to calculate holiday entitlement that accrues as hours are worked. The holiday entitlement of 5.6 weeks is equivalent to 12.07% of hours worked over a year. Therefore, it is possible to calculate an hourly rate including holiday pay by increasing the hourly rate (must be at least the NMW) by 12.07%. Full details are in the ACAS guide 'Holidays and Holiday Pay'.

Employment Legislation

General guidance and support is widely available to small businesses on employment legislation and compliance. Negligence in this area could result in an employment tribunal, so it is worth considering the benefits of subscribing to a support service such as the NFU Employment Service, which is specific to agriculture. The NFU includes a Statement of Terms and Conditions in their range of model agreements. The employer must provide a written statement within two months of the start of employment; this is not an employment contract but must include the main conditions of employment. Full details can be found at www.gov.uk. The ACAS Helpline provides free and impartial advice for employers, employees and representatives on a range of employment relations, employment rights, HR and management issues, see www.acas.org.uk. It is also possible to sign up for the monthly ACAS e-newsletter.

Redundancy

Staff who have been employed for more than two years are entitled to statutory redundancy pay that is free of tax and NIC up to a certain limit set by HMRC. Employers must be careful that they are acting fairly. It must be established that the redundancy situation is genuine and that the consultation process is adequate. For more information and advice on redundancy, visit www.acas.org.uk.

Pensions and Auto-Enrolment

The Pensions Regulator (TPR) is the UK regulator of work-based pension schemes. From 2012, all UK employers with at least one worker must provide them with access to an approved pension scheme, automatically enrol certain employees and pay contributions. It is a legal requirement and failure to comply may result in penalties. The legislation is being phased in; employers are informed of the date they must start to operate an auto-enrolment scheme, this is known as the 'staging date'. Visit www.tpr.gov.uk for comprehensive information and details of government-backed workplace pension schemes. This is a complex area and it may be worth seeking specialist advice to help select the most suitable scheme. Many existing company pension schemes do not qualify for auto-enrolment due to the low level of charges permitted. The government has set up the National Employment Savings Trust (NEST, www.nestpensions.org.uk), which must accept all employers that ask to join it. There is an upper limit on annual contributions and there are restrictions on what can be done with the fund. Currently all employees aged 22 and over, but less than state pension age (65), who earn more than £10,000 per year and normally work in the UK, must be auto-enrolled; however, there are exceptions; for example, company directors without a contract of employment. Use the 'Duties Checker' on the TPR website to find out what steps need to be taken. Following auto-enrolment, employees may wish to opt-out of the scheme. Employers must ensure that employees are not influenced in their decisions with regard to participation in the scheme, therefore good communications with employees is important. An online declaration of compliance must be completed within five months after staging date (including employers who don't have to provide a pension scheme), failure to do so may result in fines. There are also ongoing duties towards employees: paying money into their schemes, and general administration. More details can be found at www.tpr.gov.uk.

A review of the payroll administration system is recommended a month or two before the beginning of the tax year. Some payroll software incorporates a pension calculator with a facility to upload payments to the pension providers online portal. A flexible direct debit is used to collect pension payments from the employer. HMRC Basic Tools does not currently incorporate a pension calculator but there are commercial cloud-based payroll systems (HMRC approved), which do. It is also possible to enter payments manually via the pension providers website. Another useful source of information for individuals is the Money Advice Service, offering free and impartial advice. It is an independent service, set up by the government, www.moneyadviceservice.org.uk.

Health and Safety Legislation

The main legislation that applies to employment is the Health and Safety at Work Act 1974 (HSaWA) and supplementary regulations relating to employees and the workplace. This legislation is far-reaching and the structure and scope of a business will determine what legislation is applicable. Again, guidance and support is available and a service provider specialising in agriculture is recommended. For do-it-yourself situations, a comprehensive guide called 'Farmwise' is available free from the Health and Safety Executive (HSE, www.hse.gov.uk).

Health and Safety – Some Basic Essentials

An up-to-date Health and Safety Law poster together with a current employer's liability insurance certificate must be displayed in an appropriate place where staff will see it. If there are five or more employees, there must also be a written policy on health and safety. More information is available at www.hse.gov.uk.

Training Required by Law

Training in the operation and use of some agricultural vehicles and machinery is required by law, as are some courses concerned with the safety and welfare of staff. Membership of an agricultural training group is recommended. They organise farm-based training for local groups of farmers, often arranged during the winter months. More general information on training in agriculture is available on the LANTRA website, www.lantra.co.uk.

Timesheets and Working Time Regulations

The Working Time Regulations (WTR) are enforced by the HSE. It is possible to apply a derogation to the 48-hour weekly limit for adult workers, but it is not an automatic derogation; it is the characteristics of a worker's activity that determine whether or not derogation may be used. The regulations also cover rest breaks and again there is a degree of flexibility in certain circumstances. The NFU publishes a useful guide covering WTR for agricultural workers. Employees must record their daily hours of work on a timesheet. Diagram 13.1 shows a basic weekly timesheet, where more detail is required, such as work activity etc., cloud-based spreadsheets accessible via mobile phones and computers are a smarter alternative. Overtime hours and absence (holiday, sickness, etc.) should also be recorded. This information will be required to calculate gross pay and should be summarised on the timesheet.

Diagram 13.1 shows an example of a timesheet for a full-time (five days/week) agricultural worker that is paid weekly.

Week w/c	Mon	Tues	Wed	Thurs	Fri	Sat	Sun	Total
30/05/20xx	BH 8	Holiday 8	8	10	10	4	0	48
Summary for payroll	8	8	8	8	7		Basic	39
	0	0	0	2	3	4	Overtime	9
							Holiday	1

Diagram 13.1

Students and Workers from Overseas

There are civil penalties for employing illegal workers. Before a worker is taken on, the 'Legal Right to Work in the UK' check must be undertaken. It involves checking documentation, such as a passport, in the presence of the potential employee and retaining copies. An employer's guide to right to work checks can be downloaded from www.gov.uk.

All of the legislation mentioned above applies to students and workers from overseas. Leaflets containing practical information in various languages for overseas workers are available from HSE, as is further information on the HSE website, agricultural sector.

Class 1 National Insurance Contributions (NIC) will apply where employees, over the age of 25, have gross earnings above the National Insurance lower earnings limit. The employee will need a National Insurance number: details of how to apply can be found at www.gov.uk, a temporary number may be assigned when a worker starts, this is usually

Photo 13.1

automatically generated by the payroll system. Income tax will also be payable when the employee's personal tax allowance is reached. However, tax is not deducted where a 'holiday-work' student (UK or overseas) is employed, and the limit is not reached.

Piece Work

Staff picking fruit or on vegetable production lines may be paid on a 'piece-rate' basis, i.e., they are paid for the number of items that they produce or the weight of fruit picked. The Working Time Regulations apply to piece-workers, and employers must ensure that employees are paid not less than the NMW: this may mean topping up pay so that they are paid a minimum hourly rate for the hours worked.

Daily Casuals

The term 'daily casual' generally applies to beaters and casual harvest workers, where they are employed for a day, paid in cash at the end of the day, with no promise of future work. A record must be kept of each person paid including: full name, date of birth, gender, NI number, address and amount paid. This should be reported to HMRC via the payroll. If the business does not run a payroll, these records must be kept for three years after the end of the current tax year. Full details are at www.gov.uk.

Domestic Employees

The Simplified PAYE Deduction Scheme closed in April 2013. If this is the employee's only job and they are paid less than the current lower earnings limit, it is not necessary to operate PAYE at all. If they earn more than the lower earnings limit or have more than one job, then standard PAYE rules apply.

Self-Employed Individuals Providing Labour

Self-employed labour-only providers must be registered as self-employed and take responsibility for their own income tax and national insurance

Photo 13.2

contributions. Should an HMRC inspector consider that the person is not self-employed, NIC and tax would be calculated on the sums paid, which would then be payable to HMRC by the employer. Guidance on self-employed status can be found at www.gov.uk.

Agencies and Other Companies Providing Labour

Where several workers are required, rather than undertaking the task of recruiting and employing the workers directly, many farmers use agencies specialising in the provision of seasonal agricultural labour. It is important to select an agency of good reputation and compliant with Employment Agency Regulations. More information can be found about agency workers and employment agencies at www.nidirect.gov.uk.

Labour suppliers in the agricultural and related industries are regulated by the Gangmasters Licensing Authority (GLA). It is the duty of the business using labour services provided by a third party to check that they are registered with the GLA; this can be done on their website, www.gla.gov.uk. There are some exclusions particularly relevant to small farming businesses, these can also be found on the website.

Obligations of an Employer

Just as with the VAT scheme, under which every registered business acts as tax administrator on behalf of HMRC, so every employer must, without exception, register with HMRC in order to operate income-tax and national insurance schemes.

- Employees' income tax is administered as Pay As You Earn (PAYE). The employer is notified of each employee's tax code: that determines the amount that can be earned tax-free with any remainder having tax deducted at source, by the employer, for forwarding later to HMRC. Calculations are cumulative, i.e., sums are worked out on a so-far-this-tax-year basis, and refunds are possible.
- For any employee earning more than the lower earnings limit (LEL) in any particular pay period, Class 1 National Insurance Contributions are deducted from those earnings. Such sums are later forwarded, together with employer's corresponding contributions, to HMRC. Each pay period is treated in isolation; refunds cannot occur.

Photo 13.3

- Student loans repayments may also be collected via the payroll.
- The employer is also responsible for statutory payments such as sick pay, maternity pay, paternity and adoption pay.
- Employees may also be entitled to redundancy pay and it is the duty of the employer to pay at least the minimum required by law.

Registration with HMRC for a new payroll scheme is done online at www.gov.uk. It is a step-by-step process, including an option to also register with the Government Gateway; VAT-registered businesses will already have the Government Gateway User ID and password essential for all payroll systems to facilitate electronic transfer of data to HMRC.

Online Payroll Administration in Real Time

Payroll processing must be done regularly to ensure that employees receive their pay on time and an online submission must be made to HMRC. This is known as Real Time Information (RTI); the majority of employers must provide HMRC with details of payments made to employees, together with the tax and national insurance deductions, when or before the payments are made to the employee.

In the days of manual payroll processing, paper-based guides were revised and posted to all employers each year, together with revised National Insurance Tables and Employee Tax Codes. Now all the necessary up-to-date information needed to process a payroll can be found online at www.gov.uk, with the option to download and print if required.

Choice of Payroll Processing System

There are different methods of payroll processing. Almost all businesses must submit payroll information online in real time, therefore the choice is either using computerised payroll processing system or outsourcing the operation to a bureau. HMRC's 'Basic PAYE Tools' is free computer payroll software, downloadable from www.gov.uk. There are many commercial payroll software and cloud-based systems available, with varying degrees of capability and cost. Important points to consider are: broadband speed, number of employees, built-in facility for auto-enrolment and reliable software support. Outsourcing the payroll may be a cost-effective option. Many accountants and bookkeepers offer payroll processing, some with specialist agricultural knowledge.

Taking on a New Employee

The appointment of a new member of staff should be confirmed in writing, stating the start date, rates of pay, hours of work and any other information that may be necessary. Once it has been established that the person is legally entitled to work in the UK a Starter Check List (downloadable from www.gov.uk), may be used to gather most of the information needed to add a new employee to the payroll. The Starter Check List replaces Form P46 and includes the selection of an employee statement, A, B or C, which will determine the tax code to be used.

Alternatively a new employee may produce a Form P45 issued by their previous employer; this will show personal details, plus the tax code in use when they left.

Basic information required from a new employee is:

- Full name
- Address, telephone, mobile number and email address
- Date of birth and gender
- National Insurance number
- Employment start date
- Do they have a student loan?
- Bank name, sort code, account name and account number
- Contact details in case of emergency.

In addition, check and keep copies of driving licences on file, plus passports for non-UK nationals.

Payroll Preparation

It is good practice to have timesheets checked by the farmer or manager before they are passed to the administrator for summarising. The hours worked are used to calculate the gross pay for the PAYE period. Also have to hand any other information that may have arrived since the last payroll run, such as notification of a new tax code or new employee details. Management and office staff are usually paid monthly on a salaried basis.

Running the Payroll

Wages or salaries are usually agreed to be paid for a set period throughout the year. These cannot be changed without first consulting employees.

- Weekly – 52 periods (exceptionally 53)
- Two weekly – 26 periods
- Four weekly – 13 periods
- Monthly – 12 periods based on the calendar months.

It is important to check that the correct processing date and pay period is set before entering the gross pay.

The following payments made to employees are all to be regarded as taxable income for tax and NIC purposes:

- Salary and wages
- Overtime and shift pay
- Tips – unless these are paid directly to the employee or there is an independent Tronc system in place for tips (usually found in restaurants)
- Bonuses and commission
- Certain expense allowances paid in cash
- Statutory sick pay
- Statutory maternity, paternity or adoption pay
- Non-cash items such as vouchers, shares or premium bonds – PAYE is applied to the cash value of items like this.

As well as deducting income tax and NICs from employees' pay each pay period, the PAYE system may also be used to deduct or add other items such as:

- Student loan repayments
- Employees' pension contributions
- Payments under an attachment of earnings order
- Repayment of a loan you've made to an employee
- Reimbursement of expenses for which the employee has an agreement or has provided a receipt.

A payslip must be given to the employee with their pay and must show:

- Employee's gross pay (before any deductions are made)
- All deductions and the purposes for which they are made
- The net amount payable after the deductions have been made (also known as take-home pay).

The following example (Diagram 13.2) includes detail generally included on the payslip.

Employer: J & M Field			Tax Week Number 10	Tax Code 1100L	Employees Name: Mr A Driver National Insurance No AB123456 C	
	Hours	Rate	Gross		Total Gross Pay to date	4600.00
Basic	39	9.50	370.50		Total Tax Tax Paid to date	612.40
Overtime	15	13.32	199.80		Holiday Taken to date	10
Gross Pay			**570.30**			
Deductions:-					Post Tax & NI Deduction:-	
PAYE Tax			-71.75		NEST Pension	-10.00
National Insurance			-57.17			
			-128.92			
Payment Date	**05/06/20xx**		**Payment method – BACS**		**Net Pay**	**431.38**

Diagram 13.2

Income Tax and Tax Codes

Tax codes are usually made up of one or more numbers and a letter as shown on the payslip in Diagram 13.2. The number indicates the amount of pay an employee is allowed to earn in a tax year before tax becomes payable. For example, a tax code of 1100L signifies earnings of £11,000 in the current tax year before any liability to pay tax. This is referred to as free pay and, where pay is calculated on a cumulative basis, this free pay allowance is spread evenly throughout the year. If a tax code is followed by week1/month1 or an X, it indicates that should be calculated as if it is the first week/month of the tax year rather than using the usual cumulative system.

The letter following the numerical part of the tax code indicates how the tax code is adjusted to take account of any budget changes. Where an employee has more than one job, their tax free pay allowance will be allocated to their principal job and the tax code used for the second job will be BR, Basic Rate.

National Insurance Contributions (NICs)

NICs are made up of two elements:

1. Employee contributions – deducted from employee's pay.
2. Employer contributions – paid by the employer.

There are different categories of NICs. Most employees over the age of 25 pay Class 1 contributions. Class 1A contributions are payable on benefits supplied by employers to employees and directors earning more than the annual limit set by HMRC. Unlike tax, which is normally calculated on a cumulative basis, NICs are due when gross pay reaches the LEL in any pay period.

Statutory Payments to Employees

Commercial software and HMRC's Basic PAYE Tools will calculate statutory payments such as SSP or SMP and any tax and NIC due. In some circumstances, an employee may not be eligible to receive these benefits. Help and guidance is available from HMRC or payroll software suppliers and should be consulted if in doubt.

Statutory Sick Pay (SSP)

Employees who are off sick for more than four consecutive work days, are entitled to receive SSP. The first three days of sickness are 'waiting days' and do not qualify for SSP. Some businesses will continue to pay employees for short periods of sickness as part of good employment practice. Employees taken on pre 30 September 2013 with rights under the AWB may also be entitled to Agricultural Sick Pay, an enhanced variant of SSP. It is no longer possible for the employer to recover SSP from HMRC.

Statutory Maternity Pay (SMP), Statutory Paternity Pay (SPP) and Statutory Adoption Pay (SAP)

These are available to employees but there are qualifying conditions that should be checked. Employers can usually recover a proportion of these statutory payments from HMRC; this depends on the size of the annual NICs liability for the business. Small employers may qualify for Small Employers' Relief and be able to recover just over 100% of the payments.

Making PAYE Payments to HMRC and Deducting the Employment Allowance

Each month employers must pay HMRC the total PAYE summarised on Form P32 (generated by the payroll programme) by the deadline. However, small employers may be able to make quarterly payments if total deductions are below a certain limit. For deadlines, limits and methods of payment check www.gov.uk.

Within the past few years, the government introduced the Employment Allowance to reduce the cost, particularly to SMEs, of employers' national insurance and boost employment in the UK. Generally, payroll software deals with the pro-rata reduction to the payment made to HMRC each month automatically. It is important to note that where there are connecting businesses with separate payrolls, only one allowance may be claimed.

Payroll Year-End Processing

After the pay period ending on 5 April (month 12 or week 52/53) has been run, the payroll year-end must be processed. Guidance on the tasks and deadlines involved plus software updates are issued by computer payroll software suppliers. A *Certificate of Employee's Earnings, P60* must be issued to employees by 31 May following the tax-year end.

The procedure for starting the new tax year as recommended by the software provider should be followed and tax codes for the new tax year issued by HMRC should be applied from 6 April, the beginning of the new tax year.

Reporting Expenses and Benefits paid during the Tax Year

Class 1A NI and income tax is charged on various benefits provided to company directors and employees earning above a certain level of annual gross pay. This task must be completed by 6 July following the end of the tax year, preferably online, or forms may be downloaded from www.gov.uk, completed and posted.

- **P11D:** Used to report payments and benefits (for example, company cars, healthcare insurance, travel and entertainment and childcare) provided to company directors or employees at the end of the tax year. A copy must also be given to the director or employee.
- **P11D(b):** Used to report the amount of Class 1A NIC due, and declare that forms P11D have been completed. The payment must reach HMRC by 22 July.

Following the submission of P11D, income tax will be collected by adjusting the tax code either by reducing the tax-free allowance, or increasing tax due by issuing a code prefixed by the letter 'K'. An alternative to using forms P11D is to collect tax due on benefits by using the Payrolling Benefits in Kind (PBIK) service, www.gov.uk.

Calculating the expenses and benefits enjoyed by salaried directors of a farming businesses can be complicated and maybe an annual task for the accountant.

Analysis of Wages and PAYE in the Accounts

Regular labour is usually treated as an overhead cost and for a small business it may be sufficient to analyse net payments to employees and PAYE paid to HMRC under 'Wages'. This column (heading code) should be reconciled at the end of the financial year with the payroll records. In a business with a number of employees, more detailed information may be required. Departments or search codes may be used to break down 'Wages' into management, farm labour, property labour, etc. PAYE payments or refunds will also have to be apportioned and departmentalised in a similar way.

With full-cost accounts, enterprise workers are costed to the departments with which they are concerned, using detailed timesheets in the case of general and shared workers – perhaps even up to management level. More usually, regular labour is treated as a fixed cost or overhead since minor changes of emphasis between enterprises are not associated with proportionate increases or decreases in labour costs. Those labour costs, which are not only enterprise-specific but which tend to be proportional to the size of the enterprise, are treated as variable costs in any gross margin calculations; examples might include casual labour for rogueing or apple-picking. Such costs must therefore be allocated to their enterprises in the computer coding system or account analysis columns.

Filing and Security of Information

It is important to keep information held on employees secure. Payroll reports and personnel files should be kept in a locked filing cabinet and only retained for as long as they are needed. The current payroll records plus those for the preceding three years must be kept. Employers have legal obligations under the Data Protection Act; more information can be found on the website, www.gov.uk.

Top Tips

- Cloud-based payroll systems are inexpensive, safe and easy to use on any computer anywhere with a broadband connection.
- Check driving licences and passports (workers from overseas) and keep copies on file.
- Use passwords for payroll software and take regular backups.
- File staff records and payroll information in a locked drawer, make sure they are kept up-to-date and only retain for as long as it is needed.
- When in doubt, use helplines and seek specialist advice.

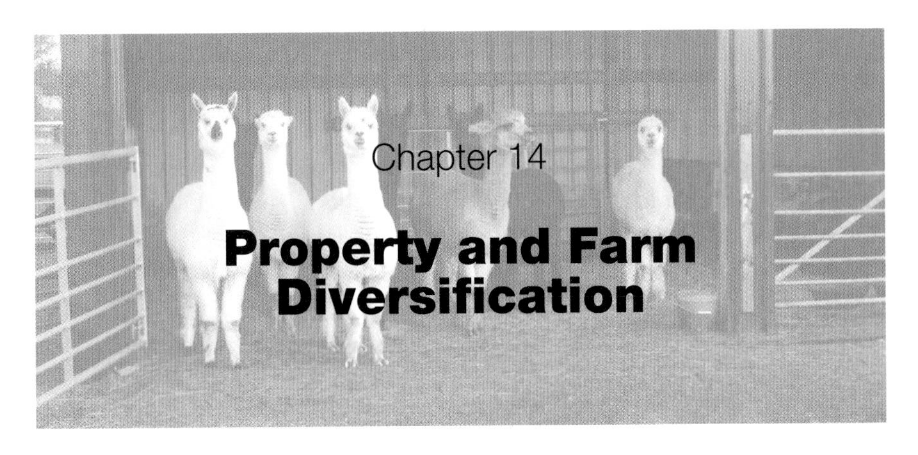

Chapter 14

Property and Farm Diversification

Property and diversified enterprises are areas that many farm administrators will be involved with at varying levels, depending on the size and nature of the business. It may be simply analysing income and expenditure on farm buildings in the farm accounts or it could be dealing with much more than the accounts, where the property lets are an important element of farm diversification.

In many large rural businesses, the services of land agents and solicitors will be used for matters connected with land and property. They are generally good at raising awareness of the benefits of managing assets properly and of the consequences of failing to do so.

Many rural businesses depend upon income from letting cottages, redundant agricultural buildings and paddocks. This income and associated expenditure may be treated as a separate enterprise within the farming business. An accounting system will need to be in place for collecting rents. Beyond financial records, there are various legal obligations and important dates to be diarised for action; it may be beneficial to use a computerised system specifically designed for rural property management alongside the accounting program.

As well as financial and accounting implications, the letting of land or buildings for non-agricultural use can often raise other important issues. Useful general guidance is available from membership organisations, such as the NFU (www.nfuonline.com) and the CLA, (www.cla.org.uk), which provide specimen model agreements. The equivalent bodies in Scotland are: NFUS and Scottish Land & Estates (www.nfus.org.uk) and (www.scottishlandandestates.co.uk). For Wales see NFU Cymru (www.nfu-cymru.org.uk). Professional advice should be sought at an early stage before embarking on any farm building conversion project; planning constraints must be taken into account. Local council websites provide useful maps and information on planning policy. There can be taxation implications affecting capital taxation or VAT or both.

Farmhouse and Buildings

A traditional family farm includes the farmhouse occupied by the farmer and family. But as well as its private residential function, that house also serves a number of business functions; these may include the farm office with its telephone, laundering of staff overalls, farm-gate sale facilities and many more. As such, a proportion of the running costs involved may be allocated to the business with the remainder constituting the 'private proportion'. There are no standards: the deemed proportion is agreed between the farmer (or the accountant) and HMRC in each individual case, to be applicable for both income tax and VAT purposes. Costs that can be allocated may include telephone, heating, power, insurance, repairs and maintenance, council tax, and writing-down allowances on shared capital assets.

General repairs and renewals to farm buildings may be treated as overheads but expenditure on improvements will increase the value of the property and needs to be reflected in the balance sheet.

Farm Cottages – and Tied Accommodation

When housing is provided for a farm worker an Assured Shorthold Tenancy (AST) can be used, but the rent must be at least £250 per year and a notice must be served on the tenant prior to the start of the tenancy. A guide (Agricultural Lettings – Housing) covering this complex area of law may be downloaded from www.gov.uk. Agricultural employees and former employees may qualify for protected occupancy; this is also explained in the guide. If the worker is required to live on the farm because of the nature of the job, council tax paid by the employer may be treated as a business cost. Overheads such as repairs and maintenance are treated as general farm expenditure and VAT may be reclaimed in full.

There is a limit to how much rent may be charged and this is set by the Valuation Office Agency – more details can be found on www.voa.gov.uk. In Scotland, if a house is provided for a farm worker, it is not subject to rent and therefore is not subject to a tenancy agreement.

Domestic Letting

An Assured Shorthold Tenancy (AST) is generally used when letting a domestic property. Farming businesses will often use the services of an agent or solicitor to set up a tenancy and deal with other legal obligations on behalf of the landlord. Agencies offer various levels of service, from finding tenants and drawing up tenancy agreements to a full property management service, including inspections and rent collection. An advantage of using a specialist for property matters is that they should be up-to-date with legislation, which is subject to change. For example, deposits must be held by Deposit Protection Service (DPS, www.depositprotection.com). An Energy Performance Certificate (EPC, www.gov.uk) is required and, if the property has a gas supply, a safety certificate is also required by law. The Association of Residential Letting Agents (ARLA) has some useful information on their website, www.arla.co.uk. Legislation in Scotland is different and more information can be found on the website of the equivalent Scottish body, Scottish Landlords, at www.scottishlandlords.com. More general information on Scottish law is available at www.scottishlaw.org.uk.

Where property letting is treated as a separate enterprise in the farm accounts, costs that can be allocated associated with the let property may be treated as variable costs rather than overheads, as would be the case for farm buildings. It may then be useful to use additional coding (departmental or search codes) for each building, so that detailed gross margins may be produced. See also Chapter 12.

Insurers should be informed when a property is let and when a property is empty. There are advantages in having a property owner's (landlord's) insurance, cover is more comprehensive at little or no extra cost.

Rent (excluding holiday lets) is usually treated as exempt for VAT purposes; this has relevance to the 'partial exemption' threshold affecting the whole business. See partial exemption below and more information in Chapter 7.

Storage Licence Agreements for Private Individuals

Some income may be generated from redundant farm buildings by offering storage facilities to private individuals in return for a periodic payment. Where the arrangement is of a private nature, for example, storage of domestic furniture by a private individual for a short period, a simple document formalising the arrangements may be drawn up. Further information and model agreements are available from the trade associations mentioned above. The important consideration here is that it is a private arrangement. If it is a trader or business wishing to make use of storage facilities, then a lease would be necessary and professional advice should be obtained. VAT is chargeable on storage.

Commercial Lets

The services of a land agent or solicitor are generally used for commercial lettings. The letting of land or redundant farm buildings for non-agricultural purposes is likely to be covered by the Landlord and Tenant Act 1954. It may also be necessary to seek planning permission for change of use.

Careful consideration should be given to any additional clauses, which may need to be included in the lease to accommodate specific needs of a prospective tenant. For example, an agricultural contractor may want to store fuel and wash down vehicles. This could have environmental implications and the responsibilities of the landlord and the tenant need to be clarified.

When the lease has been set up – as with domestic lettings – a system needs to be in place to keep track of rent collection and also dates of future rent increases and expiry of the lease. If there are a number of let properties, it may be useful to identify each property in the accounts and allocate costs to each property as described above for domestic properties.

Rent deposits should be identified within the balance sheet headings and rent deposit receipt letters kept on file. They do not have to be lodged with the Deposit Protection Service (DPS) as deposits on domestic properties do.

Improvements and Dilapidations

The farm may be held freehold or on a tenancy; in its subsidiary enterprises it may act as landlord in respect of more minor lettings of land or buildings or both. Administrators need to be aware of the existence of a vast body of relevant statutory law and custom. The general principles involved derive from common justice: it is possible for a tenant to carry out improvements to a property with the agreement of the landlord. In the event of the tenancy being terminated in the normal way, the outgoing tenant may be eligible to receive compensation (possibly based on written-down value) from the landlord. But it will be unwise to allow or to undertake any such improvements without having first committed to paper a professionally drafted agreement, unless such matters are fully covered in the tenancy agreement. Conversely, the value of a landlord's capital may be diminished by a tenant's failure to safeguard and maintain the property in accordance with the tenancy agreement; this situation can give rise to a landlord's claim on the outgoing tenant for 'dilapidations'.

Buildings and the built environment are normally to be protected against fire, damage, vandalism and the like, as well as being subjected to normal periodic maintenance such as cleaning and repainting. Basically, when the tenant leaves, for whatever reason, the property should be in such a state that it can sensibly be re-let straight away. Any landlord's expenses in reinstating the property constitute the claim for dilapidations against the outgoing tenant.

Land, particularly agricultural land, can and usually does, have secondary considerations embodied in law. Responsibility for fences and gates, water troughs and the like will have been specified in the tenancy agreement. But the tenant also has a statutory obligation to safeguard the productive resource. Failure to abide by the principles of good husbandry in respect of soil fertility, drainage and weed control, for example, can lead to disputable claims for dilapidations.

Administrators are well-advised to seek professional guidance on such matters – at the time of initiation rather than termination.

Pony Paddocks and Grazing Agreements

As with storage licences, the important consideration is whether the arrangement is with a private individual for non-business use. If this is the case, then it is possible to use a simple annual licence or agreement formalising the terms. However, if the arrangement is with a business, it is likely to be covered by the Landlord and Tenant Act 1954 and professional advice should be sought.

Photo 14.1

Graziers and Grazing Licences

Grazing licences are commonly used arrangements, setting out the terms between a landowner and grazier requiring pasture for sheep or cattle. Model agreements are available but a grazing licence is a relatively inexpensive document to have drawn up by a land agent specialising in this area. They will be aware of the need for cross-compliance (see Chapter 12). Severe poaching of grassland or grazing outside time limits by a grazier may result in a breach of cross-compliance. These are matters for careful consideration and can be covered by the licence. Public liability insurance is a responsibility of the livestock owner, who also has a duty to keep the stock securely fenced in.

Farm Tenancies

Farm Business Tenancies (FBTs) were introduced by the Agricultural Tenancies Act 1995 to encourage more letting of land by allowing greater flexibility in the farm rental market. The parties may agree some of the terms such as the rent payable, but where certain terms aren't specified in the FBT, statutory provisions within the act will apply.

In Scotland, Limited Duration Tenancies (LDTs) and Short Limited Duration Tenancies (SLDTs) were introduced by the Agricultural Holdings (Scotland) Act 2003. They are different from FBTs and are applicable in Scotland only.

This is definitely an area where professional advice should be sought.

Land and Buildings Transactions

The purchase of land and buildings, or the construction of new buildings, are capital items and should be analysed in the accounts as such. When it comes to VAT on these transactions, typically there isn't a one-rule-fits-all approach and it pays to take professional advice. Mistakes made with the VAT treatment of property are not easy to correct after the event, so the cost of obtaining advice in advance is likely to be minor compared to the potential cost of any error.

The default treatment of a land/building transaction is exemption from VAT, but there are many exceptions. Although VAT is not accounted for on a sale or rental proceeds, partial exemption rules apply, restricting the recovering of VAT on inputs such as legal fees (see below).

Holiday accommodation is classed as a taxable supply. VAT is due on the rent or sale proceeds at the standard rate.

Partial Exemption

Partial exemption may, subject to certain thresholds, apply to farming businesses where some of the outputs are taxable at the standard or zero rate and some are exempt, such as rent.

There are various methods of calculating how much input tax may be recovered against exempt income, possibly depending on the values of taxable outputs compared with exempt outputs. The rules covering partial exemption are complex and where letting is an important part of the business, the ability to deal with partial exemption may be an important consideration when selecting an accounting package (HMRC Notice 706 Partial Exemption). See Chapter 7.

De Minimis Limits in VAT

The thresholds involved in partial exemption legislation are known as '*de minimis* limits'. A farm letting out a couple of paddocks for any purpose – agricultural or otherwise – is unlikely to reach the relevant figures. However, it is the responsibility of the business to verify that it is *not* exceeding those limits. Guidance is available at www.gov.uk.

Option to Charge VAT on Rental Income

It is possible to exercise an option to tax individual commercial buildings or pieces of land, and charge standard rate VAT on the rental income thus making it possible to recover all of the VAT on expenditure for that particular unit. The terms of this option are subject to revision and up-to-date advice should be sought from an accountant or HMRC.

Photo 14.2

Other Property and Related Topics

Footpaths and Boundaries

District council websites have useful information about public rights of way. The parish council usually holds the definitive footpath map for the parish but the district council, which also holds the definitive map, deals with any legal issues concerning access and routes. General guidance on land boundaries can be found in the Land Registry section on www.gov.uk.

Cross Compliance – Good Agricultural and Environmental Conditions (GAECs)

Where footpaths, bridleways and byways go through farmland, cross-compliance applies, (GAEC 7b), they must be kept open and accessible. Similarly, boundary features such as hedgerows, stone walls etc., should be protected, (GAEC a) and trees (GAEC 7c). Full details can be found in the Guide to Cross Compliance in England at www.gov.uk; see also Chapter 12.

Maps and Utilities

There will be a variety of maps in the farm office. It is useful to have a large-scale map on the farm office wall, showing field names and numbers, location of buildings, roads and footpaths. A map showing directions to the farm is also useful.

Farmers will have copies of maps used to support the Basic Payment Scheme (BPS) claims. These maps will have been produced by the Rural Payments Agency (RPA), which have been working towards a definitive map, the Rural Land Register (RLR) in digital format, showing who is farming what in England. They are also used for Natural England and Forestry Commission schemes, Countryside Stewardship Scheme, and Farm Woodland Schemes. The Scottish Government Rural Payments and Inspections Directorate (SGRPID) provides every farmer with a map to support single-farm payment claims, Land Management Options, Scottish Rural Development Programme, Less Favoured Area Support Schemes, Forestry grant schemes, etc. For the purposes of compliance with rules affecting the above schemes, maps showing

Nitrate Vulnerable Zone (NVZ) boundaries are available online at www.gov.uk.

Land drainage maps will be needed from time to time if problems arise with field drains. They will have been provided by the drainage contractor when the drains were originally laid and consequently they may be quite old. Similarly, when gas, oil, fuel, water pipelines and underground cables are laid, maps showing the easement plan (rights of access to a third party), and the route are supplied to the landowner; these should be kept and always referred to before commencing ground excavations or building works.

There are unlikely to be official maps showing domestic water pipes, sewers, electricity cables and phone lines, but information gathered from people who have been associated with the farm for a long time can be useful. It is also useful to have a plan or at least a list of locations of electric, gas and water meters and which buildings they serve. From the health and safety aspect, a similar list of locations of main electric/gas switches and water stopcocks is a good thing to have.

Other relevant maps may include soil classification, and those showing archaeological sites or environmental features such as Sites of Special Scientific Interest (SSSIs).

Postcode and Grid Reference Conversion Tools

Useful conversion tools for finding OS grid references from postcodes or postcodes from grid references can be found on the following websites: www.freemaptools.com and www.streetmap.co.uk.

Insurance and Liability in Connection with Property Ownership

Insurers will want to know about new ventures and activities, and an on-site annual review with the agent is a useful exercise. The Farm Safety Policy might usefully be reviewed at the same time.

It is important to check that adequate public liability insurance is in place. Property owners and managers have a duty of care to people on their land. There are various guides available; for example, the Countryside Code, which applies in England and Wales, sets out the responsibilities for owners, managers and visitors, see www.gov.uk. In Scotland, there is the Scottish Outdoor Access code, which can be found at www.outdooraccess-scotland.com. Another useful publication that can also be found on this website is 'Veteran Trees: A Guide to Risk and Responsibility'.

Farm Diversification

Many farms have diversified into other business activities in addition to their core farming business. Further information on different types of farm diversification is available from www gov.uk or Scottish Government Rural Issues at www.scotland.gov.uk. The CLA and NFU may also be useful contacts. The British Institute of Agricultural Consultants (BIAC) may have members with specialist knowledge in a particular area; more details can be found on their website, www.biac.co.uk.

The following examples of farm diversifications fall into two broad categories. The first group having some impact on the farm office, and the second group requiring a careful review of resources and existing systems.

Photo 14.3

Farm diversifications requiring some changes to existing accounting procedures:

Contracting – offering services to other farmers, for example, ploughing or combining

- Whole farm contracting – providing all the labour and machinery required by another farming business.
- Environmental schemes.
- Non-food crops.
- Exotic livestock enterprises.

Where a computerised accounting system is in place, a new enterprise code will be required; with an associated sales code and codes for any direct costs relating to the new enterprise. If a cashbook system is used, new cash analysis column headings will be needed.

In the case of general contracting, there will now be customers requiring invoices. The farmer or manager will need to inform the administrator when the work has been done and invoices are to be raised. There are also cross-compliance rules covering the need to include physical information on the invoice; for example: date, type, quantity and area of fertiliser spread.

The terms of 'Whole Farm' Contracting Agreements vary. In some cases, one invoice is raised to the farmer covering the work done by the contractor at the end of the 'farming year'. Alternatively, there may be two or even three invoices raised during the year: it will depend on the terms. When the financial results of the 'farming year' are confirmed by the farmer customer, or his agent, to the contractor, a second invoice will be raised by the contractor for his share of the divisible surplus (profit) as per the terms set out in the contract.

For diversifications involving crop or livestock production, self-billing invoices are likely to be raised by the customer. Payments under environmental schemes will be made by the Rural Payments Agency (RPA), as per the agreement.

Farm diversifications requiring a review of resources and existing systems include:

- Domestic and commercial property letting, as discussed above.
- Adding value to farm produce, i.e., producing cheese, flour, etc.

- Tourism, bed & breakfast, holiday lets, etc.
- Retail, farm shops, cafés, etc.

The additional accounting and administrative work required with property letting has been dealt with above. The important consideration here is whether to outsource some of this work to a specialist. This also applies to all of the diversifications in the second group, particularly where extra staff members are taken on.

Many farm diversifications develop slowly over a period of years. A roadside stall selling eggs and a few vegetables with an honesty box can grow organically into a farm shop and café, employing several members of staff. The administration requirements of this business will also grow and the following are the main areas for consideration.

The Farm Accounts

Where the accounts have been structured around an annual farming cycle, they may now need to facilitate monthly management reporting, which in turn leads to the need for a stock control system, particularly for diversification into retail. Will the existing farm administrator, possibly a family member, have the time and skills to take on the extra accounting work involved with the diversification? If the volume of invoices increases significantly, this may be a good time to split the accounting workload between data inputting (maintaining the purchase and sales ledgers) and more complex procedures such as bank reconciliation, VAT and management accounting.

Staff, Recruitment and Training

Recruitment of staff for farm shops and cafés, and in other more specialised areas, may present a challenge. Transport and the general scarcity of suitably skilled personnel in rural areas is sometimes a problem. Many businesses outsource some of the work involved in compliance with employment legislation, such as drafting contracts of employment. Membership of an agricultural training group is strongly recommended. They will be able to advise on training courses required by legislation and more general courses, such as customer care skills. See Chapter 13 for more information.

Legislation

With new activities and more staff on the farm, the responsibilities under health and safety legislation increase and there may be a whole new area of legislation to get to grips with, covering food hygiene. Every business must carry out a fire risk assessment and depending on the number of employees, one or more members of staff will need to have attended a four-day first aid course. The local training group can help with health and safety management training. However, it may be cost-effective to use professional risk management services. Once the initial risk assessments have been carried out, procedures put in place and documented, the service provider will make regular visits and ensure that the business is compliant.

Marketing and Distribution

Designing and maintaining websites, writing press releases and sales order despatching are all vital to the success of a new venture. All of these functions could be undertaken in the farm office, if staff are prepared to up-skill and diversify along with the business. The National Farmers' Retail and Marketing Association (www.farma.org.uk) is a membership organisation offering support and marketing opportunities to farm retail diversifications.

Top Tips

- Consider using property software to record and track information on let property.
- Use a computerised diary reminder system to flag up rent reviews, etc.
- Take care when offering accommodation to farm workers, this complex area is covered in Agricultural Lettings – Housing at www.gov.uk.
- Keep tabs on farm boundaries and make sure they are registered with the Rural Land Register (RLR).
- Record any useful information that older members of estate staff may have before they retire, for example, the location of underground services such as water pipes.

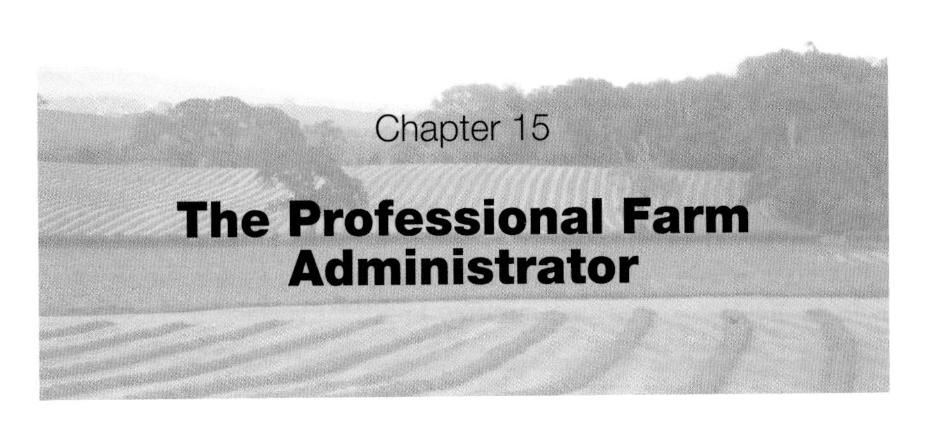

Chapter 15

The Professional Farm Administrator

In the preceding chapters of this handbook, we have looked at the diverse range of activities carried out in a farm office. In this final chapter, we consider career opportunities in this field and how to go about gathering the skills, qualifications and experience needed to become a farm administrator.

Getting Started in Farm Administration

As with many professions, it is likely that after qualifications have been gained at school and in further education, the first rung on the ladder will probably be an employed position. Many established farm administrators will have started their career with a firm of accountants, farm management consultants, agricultural cooperatives or perhaps a bank. A few years spent in a well-structured working environment can be a good foundation for a career in farm administration.

Employed or Self-Employed?

It can be a culture shock to leave the structure and support offered by a large company employing several people and become an independent professional, offering services to various small rural businesses. An important factor when considering becoming self-employed is the level of financial risk that can be undertaken. It will take time to build up a regular level of income adequate to cover outgoings, such as a mortgage and running a vehicle. Besides job security and a regular income, employment offers other perks such as pension schemes and paid holidays.

When considering setting up a mobile service, there are also the pros and cons of the location of the workplace. The convenience of towns for shopping, and possibly more social contact, may be important for some people, whereas others prefer a variety of workplaces in quiet rural locations. The types of roles, whether employed or self-employed, vary widely and may be influenced by location and the prevailing climate in agriculture, so some local research is a good idea. Talking to experienced farm administrators may be helpful; to find local contacts try the IAgSA website, www.iagsa.co.uk.

Opportunities for Employment

For those choosing the employed route, there are opportunities for qualified and experienced farm administrators on large farms and country estates. The remuneration package may include accommodation with the added advantage of avoiding any commuting and possibly living in an attractive rural location. Interesting jobs can also be found with agricultural computer software providers, in training, setting up and developing systems.

If having a variety of clients appeals, but not the responsibility of running your own business, there are positions with Land Agents and Farm Management Consultants and Bureaux that offer accounting and recording services. The type of work typically will involve travelling to clients' premises on a regular basis to undertake bookkeeping and/or recordkeeping.

Vacancies can be found in the farming press, local newspapers, recruitment agencies and websites. There is a section in the IAgSA monthly bulletin covering employment opportunities. The local branch is another good networking source available to its members.

Getting Started as a Self-Employed Farm Administrator

The process of becoming self-employed is relatively straightforward. General guidance, details of workshops on becoming self-employed and registering with HMRC may be found at www.gov.uk.

Regular contact with others doing the same sort of work is important, and joining a group offering networking opportunities early on may be useful. Sometimes personal circumstances may lead to self-employment; for example, the flexibility of working for a variety of clients on a part-time basis may fit in well with family life.

There is no easy way of establishing the demand for farm administration services in a given area at a given time. Farmers may advertise in the press or via their buying group or co-operative newsletter. They may also register their requirements with the local IAgSA branch and details will be circulated to local members. Farm administrators may promote their services using the press, buying groups or IAgSA.

There are many other rural professionals, such as accountants and land agents, who are pleased to recommend suitably qualified and experienced farm administrators to their clients. They value a professional approach to the work carried out in the farm office. It is worth spending time on selecting the right person to contact in an organisation, for example, the partner in a firm of accountants who specialises in farms. In an area of crofts or part-time holdings, informal approaches via the community pub may be productive.

Legislation for Self-Employed Farm Administrators

Keeping up-to-date with new and changing legislation is important. Guidance on general legislation affecting self-employed service providers can be found at www.gov.uk. Membership of a professional body is also helpful; it highlights new legislation relevant to its members, together with registration and compliance requirements. For example, the 2007 Money Laundering Regulations require all self-employed accounting service providers (ASPs), including professional farm administrators offering bookkeeping services, to be registered.

Filing and Security of Information

It is important to keep information held on clients secure and only retain it for as long as it is needed. Accounting service providers have legal obligations under the Data Protection Act: more information can be found on www.gov.uk/data-protection.

Insurance for Self-Employed Farm Administrators

General commercial insurance can be arranged to cover premises and equipment against fire, flood, theft and business interruption arising as a result. Public liability insurance provides cover against claims from customers and members of the public injured or adversely affected as a result of the business's activities. Employer's liability is a legal requirement as soon as staff members are employed. Contract dispute insurance, which covers a business against claims arising from disputes with clients, suppliers or third parties in relation to contractual and commercial issues, is also available. It is possible to arrange a single 'business combined policy' that can include all or some of the above-mentioned sections.

Professional indemnity insurance provides cover against claims of negligence, such as bookkeeping or recording errors. This type of cover is more specialised and can be expensive. Members of professional bodies and trade associations often have access to specialised policies at competitive rates. Third-party cover for any vehicles used for business purposes is required by law.

Selection of Clients

The advent of the Money Laundering Regulations in 2007 highlighted the importance of careful selection of clients. The regulations require that a risk assessment is carried out on new clients and reviewed annually. An introduction by an accountant of a small or medium-sized well-established farming business handling very little cash would represent a lower risk than a new business where the proprietor/s are unknown locally.

In any given rural area, there will be many connections between farming businesses and it is generally good practice to avoid taking on a new client if there is a connection with an existing client, to avoid any conflicts of interest or confidentiality. However, there are exceptions and these will depend upon the relationship between the clients and the administrator.

Starting with a New Client

During an initial contact, the prospective client will outline the nature and extent of the work involved. Where this falls within scope of the services offered by the farm administrator, details of working arrangements can be discussed.

Guidelines on working arrangements, which are designed to be a starting point, are available to IAgSA members. It is recommended that a checklist is devised to make sure that key areas of responsibilities are covered, together with details of remuneration.

In an established business with efficient systems in place and existing roles well-defined, these initial negotiations may be straightforward. However, this is rarely the case with small rural businesses. The prospective clients may be new to the concept of hiring farm office services but have chosen to do so because it is time to introduce a computerised system and they have neither the time nor skills to do it themselves. In a case like this, it may take a few visits from the administrator and discussions with the business owners to establish the level of service required and the responsibilities of each party.

It is good practice to get an *agreement of terms and services* in place as soon as possible, setting out the responsibilities of each party, including online procedures (banking, VAT and payroll), payment terms, relevant legislation and code of conduct. A model service agreement is available to IAgSA members; it is a comprehensive document devised specifically for self-employed farm administrators.

Professional Bodies and Codes of Conduct

Throughout this chapter, we highlight the benefits of belonging to a professional body. IAgSA is the only professional body for farm administrators, membership of which demonstrates a level of skills and qualifications to prospective clients. The different grades of membership and entry level requirements can be found on the IAgSA website www.IAgSA.co.uk.

There are other membership organisations, which farm administrators may find helpful. These include the Institute of Certified Bookkeepers (ICB), www.bookkeepers.org.uk and the Association of Accounting Technicians (AAT), www.aat.org.uk. Each organisation offers a range of courses and professional development opportunities.

All professional bodies have a code of conduct that members must adhere to. These codes vary but usually include the following points.

Members must:

- Preserve the confidentially of clients' affairs.
- Avoid engaging in any activities that will bring the member or the professional body into disrepute.
- Not prepare any record which they know to be false.
- Not accept work for which they have insufficient expertise.

Each of the organisations mentioned holds regular meetings within its own network of branches. Meeting other people in the same line of work, who encounter similar problems and challenges often helps to avoid professional isolation.

Skills and Qualifications

Whether employed or self-employed, the role of the farm administrator is an important and responsible one. A nationally recognised bookkeeping or accounting qualification is an excellent foundation and starting point. It is possible to complement this with the *IAgSA Rural Business Administration Programme*, a modular course covering more specialised agricultural recording, terminology and legislative compliance. Getting qualified can be expensive, but there may be regional funding available for individuals from agricultural charitable trusts, such as the Chadacre Agricultural Trust (East Anglia), which has part-funded this book. A UK-wide list of such bodies may be found at www.afcp.org.uk.

The IAgSA Mentoring Programme offers 12 months of help and advice to Associate Members who are working while completing their training. Experienced IAgSA members serve as mentors. They provide useful guidance and signposting to relevant information to facilitate professional development. See www.iagsa.co.uk website for full details of the training and mentoring programmes.

There are no mandatory qualifications required to provide farm administration services, but in reality, sound IT skills, training in accounting practices and associated software are required as well as general farm business knowledge.

Continuing Professional Development (CPD)

Professionals engaged in any area of business, including agriculture, are increasingly recognising the benefits of adopting a planned and structured approach to the maintenance and upgrading of their skills.

IAgSA offers a CPD scheme designed to acknowledge those members who are constantly upgrading their professional development, by increasing their knowledge through planned and constructive learning. Members with CPD recognition for their updated and improved skills then have an advantage in the marketplace. ICB and AAT also run similar schemes.

IAgSA's regional branches run workshops covering accounting and other topical issues affecting the industry. Their national conference provides more opportunities for CPD. It takes place each spring in different locations throughout the UK but always with a local flavour, giving delegates the chance to discover the great diversity of farming and associated administrative practices.

Agricultural Colleges and Agricultural Training Associations are generally willing to run short courses if there is sufficient demand. Agricultural software companies run regional workshops designed to keep their customers and farm administrators up to date with latest developments in technology.

HM Revenue & Customs (HMRC) and Business Gateway in Scotland offer business workshops and online webinars. Check details on their websites.

Professional Boundaries

As the agricultural business environment becomes more complex, the need to engage the services of specialists increases. It generally pays to get legal documents drawn up by a solicitor or land agent but sometimes there is the temptation to 'get it done on the cheap'. Farm administration is also specialised work and practitioners need to be aware of their limitations.

This is particularly relevant to self-employed administrators, who may be asked to take on work that is potentially outside their limit of competence. It may feel like a welcome new challenge, but there could be serious consequences if things go wrong. This is where it is useful to have an agreement of terms and conditions in place between the client and farm administrator, clearly stating the range of services provided. For the employed administrator, the job description and contract of employment serve a similar purpose.

Working with Clients and Support Organisations

Working as a farm administrator is a privileged position, in that the farmer is allowing access to the inner sanctum of his farm, i.e., the farm office, and in particular the farm finances. There will be times when it appears that all is not well and the question of mentioning any misgivings to the farmer may arise. The reaction to this depends to a large extent on the strength of the relationship with the client. However, the financial recordkeeping does fall within the administrator's sphere of operations on the farm, so it is not impertinent for them to query matters that they feel may need to be explored in more depth.

The financial side of a farm business is only one area where things can go wrong; there may be other issues: some triggers are obvious, such as a sudden illness or an accident within the family; others, such as mental illness or family break-ups, are more difficult to confront and acknowledge. Many farmers are staunchly protective of their independence and resist calling for any help. An early intervention, however, gives everyone a much better chance of finding a solution. From the administrator's position of privilege, it might be possible to persuade farmers that they need support, and that the sooner they ask for help, the better. If unsure of what to do, speak to the local representative of IAgSA or other professional body (ICB, AAT, etc.). Alternatively, other useful contacts can be found on www.farminghelp.org.uk. This is a dedicated organisation of three charities, each of which offers support in slightly different but complementary ways to farmers in distress, whatever the cause. Indeed, in some cases, they suggest the involvement of a farm administrator.

Top Tips

- A nationally recognised bookkeeping or accounting qualification is an excellent foundation for a career in farm administration.
- Join a professional body and go along to local meetings and meet others involved in the same line of business.
- Confidentiality is extremely important and the basis of a good working relationship.
- Keep a manual of office procedures, deadline and important contacts.
- Get to know other professionals involved with the business – i.e., the accountants, land agents, NFU, CLA, TFA – and work with them.

Appendix 1

Useful Contacts and Online Resources for Farm Administration

Key Membership Organisations

IAgSA	01604 770372
NFU	024 76858500
Callfirst (Members)	0370 8458458
NFU (Scotland)	0131 472 4000
NFU (Wales)	01982 554200
FUW Farmers Union of Wales	01970 820820
CLA	0207 2350511
Scottish Land & Estates	0131 653 5400
TFA	0118 9306130
TFA Scotland	01408 633275

Government Helplines:

VAT	0300 200 3700
VAT Online	0300 200 3701
PAYE Employer	0300 200 3200
Pay and Work Rights (ACAS) For Employer and Employees	0300 123 1100
Defra Helpline	03459 33 5577
Defra Rural Services	03000 200 301
Environment Agency	03708 506 506

Environment Agency	
– Water Pollution Hotline	0800 807060
– Floodline	0345 988 1188
Environment Agency Scotland	03000 99 66 99
Pollution Hotline	0800 80 70 60
HSE Infoline	0300 003 1747
Farming Connect (Wales)	08456 000 813

Rural Support Networks

Addington Fund.	01926 620135
Farming Help FNC	01788 510866 Helpline: 03000 111 999
RABI	01865 724931 Helpline: 0808 2819490
RSABI (Scotland)	0131 4724166 Helpline: 0300 1114166
Arthur Rank Centre (Germinate)	02476 853060

Primary Source of Business Information Classified by Sector, i.e. Farming or Agriculture

Business and Farming (England)	www.gov.uk
Business Info (Northern Ireland)	www.nibusinessinfo.co.uk
Business Gateway (Scotland)	www.gov.scot
Business Wales (Wales)	www.gov.wales

AHDB	Agricultural and Horticultural Development Board www.ahdb.org.uk
Includes links to:	Beef and Lamb Cereals and Oil seeds Dairy Horticulture Pork

	Potatoes MLC Services Ltd
DEFRA	Department for Environment Farming and Rural Affairs www.defra.gov.uk
DARDNI	Department of Agriculture and Rural Development Northern Ireland www.dardni.gov.uk
EA	Environment Agency www.environment-agency.gov.uk
HMRC	HM Revenue and Customs www.hmrc.gov.uk
HSE	Health and Safety Executive www.hse.gov.uk
NE	Natural England www.naturalengland.gov.uk
RPA	Rural Payments Agency www.rpa.gov.uk
SGRPID	Scottish Government Rural Payments Inspections Directorate www.ruralpayments.org

Other Sources of Useful Information for the Rural Sector

ABC	Agricultural Business Costings www.abcbooks.co.uk
ABRS	Association of British Riding Schools www.abrs-info.org
ACAS	Advisory, Conciliation and Arbitration Service www.acas.org.uk
ACRE	Action with Communities in Rural England www.acre.org.uk
AFS	Assured Food Standards www.redtractor.org.uk
AFCP	AgriFood Charities Partnership www.afcp.org.uk

AICC	Association of Independent Crop Consultants www.aicc.org.uk
AIC	Agricultural Industries Confederation www.agindustries.org.uk
ALA	Agricultural Law Association www.ala.org.uk
AMC	Agricultural Mortgage Corporation www.amconline.co.uk
ADDINGTON	Addington Fund www.addingtonfund.org.uk
ARC	The Arthur Rank Centre www.arthurrankcentre.org.uk
ARLA	Association of Residential Letting Agents www.arla.co.uk
BASC	British Association for Shooting and Conservation www.basc.org.uk
BASIS	Basis Registration Ltd www.basis-reg.co.uk
BHS	British Horse Society www.bhs.org.uk
BIAC	British Institute of Agricultural Consultants www.biac.co.uk
BCMS	British Cattle Movement Service www.bcms.gov.uk
BDS	British Deer Society www.bds.org.uk
BGS	British Goat Society www.britishgoatsociety.com
British Grassland Society	www.britishgrassland.com
British Camelids Association	www.llama.co.uk
British Growers Association	www.britishgrowers.org
British Poultry Council	www.britishpoultry.org.uk

British Wool Marketing Board
www.britishwool.org.uk

CAAV The Central Association of Agricultural Valuers
www.caav.org.uk

CAB Citizens Advice Bureau
www.citizensadvice.org.uk

CCC Camping and Caravanning Club
www.campingandcaravanningclub.co.uk

CLA Country Land and Business Association
www.cla.org.uk

Commercial Horticultural Association
www.cha-hort.com

Companies House
www.companieshouse.gov.uk

COSHH Control of Substances Hazardous To Health
www.gov.uk

Data Protection Gov.uk
www.gov.uk

DPS Deposit Protection Service
www.depositprotection.com

DVLA Driver and Vehicle Licensing Agency
www.dft.gov.uk/dvla

DWP Department for Work and Pensions
www.dwp.gov.uk

EFFP English Farming and Food Partnership
www.effp.com

FACE Farming and Countryside Education
www.face-online.org.uk

FACTS Fertiliser Advisers Certification and Training Scheme
www.factsinfo.org.uk

Family Farmers' Association
www.familyfarmersassociation.org.uk

The Farmers Club
www.thefarmersclub.com

FARMA National Farmers Retail and Markets Association
www.farma.org.uk

Farming Help	www.farminghelp.co.uk
Farm Stay UK	www.farmstay.co.uk
FCN	The Farming Community Network www.fcn.org.uk
Forestry Commission	www.forestry.gov.uk
FSA	Food Standards Agency www.food.gov.uk
FSB	Federation of Small Businesses www.fsb.org.uk
FUW	Farmers Union of Wales www.fuw.org.uk
FWAG	Farming and Wildlife Advisory Group www.fwag.org.uk
GLA	Gangmasters Licensing Authority www.gla.gov.uk
Government Gateway	www.gateway.gov.uk
GWCT	Game & Wildlife Conservation Trust www.gwct.org.uk
HCC	Hybu Cig Cymru (Meat Promotion Wales) www.hccmpw.org.uk
HDC	Horticultural Development Council – SEE AHDB Hort
IA	Innovation for Agriculture www.innovationforagriculture.org.uk
IAGRM	Institute of Agricultural Management www.iagrm.org.uk
IAgSA	Institute of Agricultural Secretaries and Administrators www.iagsa.co.uk
ICAEW	Institute of Chartered Accountants of England and Wales www.icaew.com
LANDEX	Landbased Colleges Aspiring to Excellence www.landex.org.uk
Land Registry	www.landregistry.gov.uk
LANTRA	Training Association for Landbased Sector www.lantra.co.uk

LEAF — Linking Environment and Farming
www.leafuk.org

Learndirect — www.learndirect.com

MLR — Money Laundering Regulations
www.hmrc.gov.uk

MRA — Machinery Ring Association of England and Wales
www.machineryrings.org.uk

NAC — National Apprenticeship Service
www.apprenticeships.org.uk

NAAC — National Association of Agricultural Contractors
www.naac.co.uk

NBA — National Beef Association
www.nationalbeefassociation.co.uk

NDC — National Dairy Council
www.nationaldairycouncil.org

NFU — National Farmers Union
www.nfuonline.com

NFUS — NFU Scotland
www.nfus.org.uk

NFU CYMRU — National Farmers Union Wales
www.nfu-cymru.org.uk

NI Direct — Northern Ireland Dept of Agriculture & Rural Development
www.dardni.gov.uk

NIAB — National Institute of Agricultural Botany
www.niab.com

NLW — National Living Wage
www.livingwage.gov.uk

NMW — National Minimum Wage
www.gov.uk

NPA — National Pig Association
www.npa-uk.org.uk

NPTC — National Proficiency Test Council
www.nptc.org.uk

NSA — National Sheep Association
www.nationalsheep.org.uk

NRoSO — National Register of Sprayer Operators
www.nroso.org.uk

Britains Rural Organisation and Events on the Map
www.nationalrural.org

Ordnance Survey
www.ordnancesurvey.co.uk

OF&G — Organic Farmers & Growers
www.organicfarmers.org.uk

Pension Regulator
www.thepensionsregulator.gov.uk

PSD — Pesticides Safety Directive
www.pesticides.gov.uk

Pocketbook — John Nix Farm Management Pocketbook
www.thepocketbook.biz

QMS — Quality Meat Scotland
www.qmscotland.co.uk

RABI — Royal Agricultural Benevolent Institution
www.rabi.org.uk

RASE — Royal Agricultural Society of England
www.rase.org.uk

RBR — Rural Business Research
www.fbspartnership.co.uk

RBST — Rare Breeds Survival Trust
www.rbst.org.uk

Red Tractor Assurance for Farms
www.redtractor.org.uk

REA — Renewable Energy Association (Assured British Standards)
www.r-e-a.net

RICS — Royal Institution of Chartered Surveyors
www.rics.org/uk

RLR — Rural Land Registry
www.gov.uk

RSPCA — Royal Society for the Protection Cruelty to Animals
www.rspca.org.uk

RSPB Royal Society for the Protection Birds
www.rspb.org.uk
RSABI Royal Scottish Agricultural Benevolent Institution
www.rsabi.org.uk
Royal Welsh Agricultural Society
www.rwas.co.uk
Royal Highland and Agricultural Society of Scotland
www.rhass.org.uk
RUAS Royal Ulster Agricultural Society
www.balmoralshow.co.uk
Rural Support Wales
www.ruralsupportwales.org.uk
SA Soil Association
www.soilassociation.org
SAOS Scottish Agricultural Organisation Society
www.saos.co.uk
Scottish Association of Landlords
www.scottishlandlords.com
Scottish National Heritage
www.snh.gov.uk
Scottish Law on Line
www.scottishlawonline.org.uk
The Scottish Rural Network
www.ruralnetwork.scot
Scottish Outdoor Access Code
www.outdooraccess-scotland.com.
SEARS Scotland's Environment and Rural Services
www.sears.scotland.gov.uk
SLE Scottish Land & Estates
www.scottishlandandestates.co.uk
SMRA Scottish Machinery Ring Association
www.scottishmachineryrings.co.uk
SQCrops Scottish Quality Crops
www.sqcrops.co.uk
SRUC Scotland's Rural College
www.sruc.ac.uk

SWT	Scottish Wildlife Trust www.scottishwildlifetrust.org.uk
TFA	Tenant Farmers Association www.tfa.org.uk
TFAS	Tenant Farmers Association Scotland www.tfascotland.org.uk
UK 200 Group	UK 200 Group of Accountants & Lawyers www.uk200group.co.uk
UKBA	UK Borders Agency www.ukba.homeoffice.gov.uk
UKWAS	UK Woodland Assurance Standard www.ukwas.org.uk
VMD	Veterinary Medicines Directorate www.vmd.gov.uk
VOA	Valuation Office Agency www.voa.gov.uk
VOSA	Vehicle and Operator Services Agency www.vosa.gov.uk
Wales	Farming Connect www.wales.gov.uk
Wildlife Trusts	www.wildlifetrusts.org
WiRE	Women in Rural Enterprise www.wireuk.org
Woodland Trust	www.woodlandtrust.org.uk
YFC	Young Farmers Club www.nfyfc.org.uk
YFC Scotland	Scottish Association of Young Farmers Clubs www.sayfc.org

Online Resources for Businesses

Government Gateway	www.gateway.gov.uk Registration is required to access government online services.

Online Resources for the Farming Sector

ALL available from:

www.gov.uk or www.gov.scot or www.gov.wales/rpwonline
Cattle Tracing System (CTS) Online
Basic Payment Scheme (BPS) RPA Online
Countryside Stewardship Online
Entry Level Stewardship (ELS) Online
Higher Level Stewardship (HLS) Online
English Woodland Grant Scheme

Farm Assessments

Cross Compliance Self-Assessment Tool
Soil Protection Review
Nitrate Vulnerable Zones (NVZ) Self Assessment Tool
Nitrate Vulnerable Zones (NVZ) Derogation Application
Catchment Sensitive Farming (CSF) Advisory Tool

Agricultural Surveys

June Agricultural Survey

Livestock Calculators and Market Reports

Animal Feed Registration Tool – FSA
Livestock Movement Licence
English lamb and beef market reports – AHDB
Livestock farm business calculators

Other Business Tools

Food Standards Agency (FSA) guidance tool
Farm Business Survey benchmarking online tool

Online Resources for the Farming Sector via Other Government Departments

HMRC On-line Submission

VAT, Monthly/Quarterly Returns plus Direct Debit set up.
Payroll, P45s, P46s and Year End Returns plus Direct Debit set up.
Email Alerts also available for key tax deadlines.

Health & Safety Executive

Report an incident online (RIDDOR).
Farm Self-Assessment software download.
Email Alerts available.

Environment Agency

Register agricultural waste online.

DVLA

Apply for a tax disc online, make enquiries about a vehicle and other driving related matters.

Accounting and Management Software Suppliers

Ac-uman Accounts
www.accu-man.co.uk
Accounting and property management software.

Agridata
www.agridata.co.uk
Software programs for beef, dairy and sheep enterprises.

Farm Matters Ltd
farm-software.co.uk
Farm management software for Cross Compliance, EID, Cattle, Sheep, Arable, NVZ, Vet and Medicine records.

Farmdata Management Systems
www.farmdata.co.uk

Software programs for livestock, crops, financial and payroll.

Farmplan Ltd
www.farmplan.co.uk
Financial and management software for farms, estates and rural businesses. Hardware also supplied.

Farm Solutions
www.farmsolutions.co.uk
Range of farm software for financial and physical data.

FarmWorks by Shearwell Data
www.shearwell.co.uk
Complete management program for the cattle and sheep farmer with data transfer capabilities.

Knight Tustain
www.knight-tustian.co.uk
Software in a range of clear and simple programs for farmers, growers and small businesses.

Landmark Systems
www.landmarksystems.co.uk
Simple cost-effective software for accounting and property management.

Muddy Boots www.muddyboots.com
Comprehensive, easy-to-use, farm recording for all aspects of crop production.

Pear Technology Services Ltd
www.peartechnology.co.uk
Advanced crop management, agronomy, contracting, mapping and precision farming software.

Planet Nutrient Management
www.planet4farmers.co.uk
Industry standard software to help farmers in England, Wales and Scotland with field-level nutrient management, and compliance with the Nitrate Vulnerable Zones (NVZ) regulations.

Tried and Tested Professional Nutrient Management
www.nutrient management.org

SAC Livestock Record Program

www.sruc.ac.uk

Complete management program for cattle and sheep, meeting all statutory records and management requirements.

Sum-It Computer Systems

www.sum-itsoftware.co.uk

Integrated farm management software designed to handle all types of farm and rural business enterprises – including software for farm accounts, dairy, beef, sheep and field records.

Appendix 2

Physical Performance Indicators

Basic Calculations	*What it is used for…*	*How to calculate it…*	
Fractions > Percentages	More widely used comparison than fractions.	$\frac{5 \times 100}{8} = \frac{500}{8} = 62.5\%$	
Percentage +/-	To find the percentage increase or decrease from one number to another.	From 5250 to 6750, (6750 – 5250) × 100 ÷ 5250 = 150000 ÷ 5250 = 28.6% increase From 6750 to 5250, (6750 – 5250) × 100 ÷ 6750 = 150000 ÷ 6750 = 22.2% decrease	
Ratios	Direct comparisons.	Divide the first by the second to give a ratio of the form ?:1. Or a figure by itself may imply … to one: see Food Conversion Ratio below.	
Physical performance indicators	***What it is used for…***	***How to calculate it…***	***Websites with more calculators, templates and comparisons…***
Lambing Percentage & Rearing Percentage	Measures of yield/ewe taking into account mortality – useful comparison year on year and national benchmarking.	**LP** = Lambs born / number of ewes put to the tup × 100 **RP** = Lambs reared* / number of ewes put to the tup × 100 *sold as finished/store lambs or sold/retained for breeding	The organisation for the English Beef and Sheep Meat Industry www.ahdb.org.uk Useful 'What if Calculator'
Calving Index	Measure of the regularity of breeding of cows. The average of all the most recent calving intervals. Should approximate 365 days.	From calving records, calculate the number of days (interval) for each cow then work our the average for the herd / Number of Cows	The organisation for British Dairy Industry www.ahdb.org.uk www.hccmpw.org.uk
Farrowing Index	Measure of the regularity of breeding of sows. The average of the most recent litters/sow/year. Target in the region of 2.3 litters/ sow/year.	365 Days / Sow Cycle (Example: Sow Cycle = gestation 115 days average + 25 lactation days average + weaning to service 5 days average = 145)	The organisation for British Pork www.ahdb.org.uk Calculator/comparison of key performance figures.
Killing Out Percentage	Expresses the proportion of a live animal's weight which when killed is sold as a carcase.	Deadweight × 100 / liveweight.	www.ahdb.org.uk 'Finishing Calculator'
Liveweight Gain per Animal	Needed to calculate food conversion ratio.	Finishing weight (e.g. 90 kg baconer) – Starting weight (e.g. 15kg weaner).	
Liveweight Gain per Batch	Needed to calculate food conversion ratio and rate of LWG.	Current batch weight + weight of animals sold or transferred – weight of animals transferred in – batch weight at start of period.	
Food Conversion Ratio – FCR – *in meat animals fed on concentrates.*	Kilograms of concentrate feed per kilogram LWG.	(Initial stock of feed + deliveries – stock at end of period) / LWG over the same period.	The organisation for British Pork www.ahdb.org.uk
Egg Percentage Production	Percentage of hens laying each day. Normally rises to ~95% and then declines.	Eggs laid each day × 100 / hens housed at point of lay.	

Appendix 3

Financial Terms and Key Performance Indicators

General Farm Accounting Terms	*Definition of terms*
Enterprise	Any operation that could, in principle if not in practical husbandry, be carried out separately: for example, winter wheat, early potatoes, egg production, pig breeding, cut flowers.
Output	Sales + closing valuation – opening valuation. Normally considered on a gross whole-enterprise basis, so cereal output includes any straw before levy and haulage; dairying includes culls and calves; bid prices of auctioned commodities.
Variable cost	Any cost that varies in direct proportion to the size of any proposed change in an enterprise. (It is used in gross margin analysis as a planning tool. There's no planning for 'no change': all planning concerns possible future changes.)
Fixed cost	Any cost that is not a variable cost as defined here (regardless of whether or not it can be assigned to any particular enterprise).
Gross margin	Enterprise output minus its variable costs. It measures the increased or decreased contribution that enterprise would make towards the fixed costs and profit if that enterprise were to be increased or decreased. It is also widely used in benchmarking.
Profit	Taxable profit is business profit adjusted for tax purposes, including adjustments and allowances as set out by HMRC for each tax year.
Net farm income	When used for benchmarking purposes, it represents the return to the farmer and spouse alone for their manual and managerial labour and on their tenant-type investment in the farm business, see also www.fbspartnership.co.uk. However, it may also be described as the gross farm income less cash expenses and non-cash expenses, such as capital consumption and farm household expenses.
Landlord's capital	'Capital' in farming refers to the total finance required in order to operate a farm business. In the case of an owner-occupier this includes the purchase of land with its associated facilities: roads, drains, ditches, hedges and at least most of the buildings. This is referred to as landlord's capital, fixed capital or long-term capital.
Tenant's capital	The remaining capital that is needed to equip, stock and run the farm, is normally called 'tenant's capital'.
Working capital	The cash needed to cover the day-to-day operations of the business.
Depreciation	A method of allocating the cost of a tangible asset over its useful life. **Straight line** depreciation is the simplest method, dividing the total cost of an asset by an estimate of its working life. The calculation should also have to take into account any selling or scrap value that might exist were the asset to be sold at the end of that working life.

General Farm Accounting Terms	*Definition of terms*
	The **reducing balance** method calculates the depreciation by writing off the same percentage rate each year. This method charges a lot of depreciation in the early years and less later.
Self-supply	Produce taken out of the business free of charge, such as meat, vegetables, etc. This is far less common these days than it used to be. Due to traceability, etc., there is far less opportunity to take assets for personal use.
Balance Sheet Terms	***Definition of terms***
Fixed assets	Assets that are purchased for long-term use and are not likely to be converted quickly into cash, such as land, buildings, machinery, breeding livestock (Herd basis).
Current assets	Cash and other assets that are expected to be converted to cash within a year, such as stock and debtors.
Physical working assets	Raw materials and stocks of the business likely to be converted into cash within a production cycle; for example, trading livestock, crops, consumable stocks.
Liquid assets	Effectively cash assets of the business; for example, bank balances, sundry debtors, pre-payments.
Total assets	Sum of fixed and current assets.
Liabilities	Something for which the business is responsible, for example a debt or financial obligation.
Long-term loans	Debts liable to be repaid in more than one year; bank loans, mortgages.
Current liabilities	Debts liable to be repaid within one year; bank overdraft, creditors, accruals.
Net worth (owner's equity)	Residual claim of the owners of the business after all external claims have been met.
Total liabilities	Sum of loan capital, current liabilities and net worth. (Note that these are business accounts: the business is ultimately liable to settle its debts to the owners. So owners equity is a liability, not an asset!)

Key Performance Indicators (KPIs)	*What it is used for...*	*How to calculate it...*
Gross profit Percentage	Commonly used ratio across all business sectors, which looks at gross profit as a percentage of turnover. Farming terminology is different, see above, but the sum is the same.	Gross profit/turnover x 100 Gross margin/gross output x 100
Farm business income (FBI)	A measure of comparison of farm type. Excludes rental income but includes income from diversified activities where they are included in the farm accounts.	Total output from agriculture (including valuation change) including agri-environmental schemes, diversification, single farm payment LESS variable costs and overheads plus profit/loss on sale of fixed assets.
Management and investment income (M&II)	Farm business income (FBI) less any imputed charge for farmer and spouse labour costs. Represents the reward to management and return on tenant-type capital.	FBI – cost of farmer & spouse labour
Total tenant's capital	Total assets less value of owner occupier assets (see Landlord's capital above – *owner-occupier...*) and tenant's improvements.	Usually calculated as the average of opening and closing live and dead stock valuations

Key Performance Indicators (KPIs)	***What it is used for...***	***How to calculate it...***
Return on tenant's capital	M&II as a percentage of Tenant's Capital.	M&II x 100/tenant's capital
Return on capital employed	This calculation gives an indication of the return (profit the business has made in a period) against the money that the owner has tied up in the business.	Pre-tax profit/capital employed
Current ratio	This figure is of particular interest to bankers when negotiating a loan facility, it measures company's ability to meet financial obligations. Expressed as the number of times current assets exceed current liabilities. A high ratio indicates that a company can pay its creditors as they fall due. A number less than one indicates potential cash flow problems.	Current assets/current liabilities
Borrowing ratio/Gearing ratio	A general term describing a financial ratio that compares some form of owner's equity (or capital) to borrowed funds. Gearing is a measure of financial leverage, demonstrating the degree to which a business's activities are funded by owner's funds versus creditor's funds.	Different ratios are used including: Asset ratio (total debt / total assets) and Debt-to-equity ratio (total debt / total equity)
Debtor ratio/ debtor days	Some double entry accounting programs will report this ratio automatically. It becomes important when the business has a significant number of customers. The 'debtor days' ratio indicates the efficiency of the business in collecting money from customers. Where terms are set at 30 days and this figure is more than say 45 days a review of procedures may be a good idea.	Trade Debtors / Sales X 365 = Avg collection period
Creditor ratio/ creditor days	A similar calculation to that for debtors, however not of much concern in a well run business where accounts are settled on time.	Trade Creditors / Cost of Sales x 365 = Avg payment period
Stock turnover	Stock turnover is an important monitor for a retail business (farm shop/country store) but otherwise not generally referred to in farm accounts. It helps answer questions such as 'is there too much money tied up in stock?' An increasing stock turnover figure or one which is much larger than the 'average' for retail, may indicate poor stock management.	*Cost of sales/average stock value *(*same as gross margin)*
Liquidity ratio	High figure may indicate that money is not being used effectively: low figure may cast doubt on cash flow vulnerability.	Liquid assets/current assets
Fixed asset ratio	Fixed Asset Ratio measures the ratio of fixed assets to total assets and gives an assessment of whether too much capital is tied up in fixed assets.	Fixed assets/total assets
Owner's equity ratio	Reflects the owner's personal stake in the business: high figures may encourage lenders.	Owner's equity in the business/total assets

Appendix 4

Conversions

Metrication

Weight = Mass	
tons × 1.016 = tonnes	tonnes × 0.9842 = tons
tons × 1016 = kg	kg × 0.0009842 = tons
cwt × 50.80 = kg	kg × 0.01968 = cwt
lb × 0.4536 = kg	kg × 2.205 = lb
oz × 28.35 = g	g × 0.03527 = oz

Distance = Dimension	
miles × 1.609 = km	km × 0.6214 = miles
chains × 20.12 = m	m × 0.04971 = chains
yards × 0.9144 = m	m × 1.094 = yards
feet × 0.3048 = m	m × 3.281 = feet
inches × 25.40 = mm	mm × 0.03937 = inches

Area	
sq miles × 2.590 = km^2	km^2 × 0.3861 = sq miles
acres × 0.4047 = ha	ha × 2.471 = acres
sq yards × 0.8361 = m^2	m^2 × 1.196 = sq yards
sq feet × 0.09290 = m^2	m^2 × 10.76 = sq feet
sq inches × 6.452 = cm^2	cm^2 × 0.1550 = sq inches

Volume	
cu yards × 0.7646 = m^3	m^3 × 1.308 = cu yards
cu feet × 28.32 = l	l × 0.03531 = cu feet
cu inches × 16.39 = ml	ml × 0.06102 = cu inches

Capacity

bushels × 36.37 = l	l × 0.02750 = bushels
bushels × 0.03637 = m^3	m^3 × 27.50 = bushels
gallons × 0.004546 = m^3	m^3 × 220.0 = gallons
gallons × 4.546 = l	l × 0.2200 = gallons
pints × 0.5683 = l	l × 1.760 = pints
fl oz × 28.41 = ml	ml × 0.03520 = fl oz

Rates

tons/acre × 2.511 = tonnes/ha	tonnes/ha × 0.3983 = tons/acre
tons/acre × 2511 = kg/ha	kg/ha × 0.0003983 = tons/acre
cwt/acre × 125.5 = kg/ha	kg/ha × 0.007966 = cwt/acre
plant food units/acre × 1.255 = kg/ha	kg/ha × 0.7966 = plant food units/acre
lb/acre × 1.121 = kg/ha	kg/ha × 0.8922 = lb/acre
oz/sq yd × 33.91 = g/m^2	g/m^2 × 0.02949 = oz/sq yd
gals/acre × 11.23 = l/ha	l/ha × 0.08902 = gals/acre
pints/acre × 1.404 = l/ha	l/ha × 0.7121 = pints/acre
acre-inches × 254.0 = m^3/ha	m^3/ha × 0.003937 = acre-inches

Miscellaneous

tons/cu yard × 1.329 = tonnes/m^3	tonnes/m^3 × 0.7525 = tons/cu yard
lb/cu feet × 16.02 = kg/m^3	kg/m^3 × 0.06243 = lbs/cu feet
lb/gal × 99.78 = g/l	g/l × 0.01002 = lb/gal
oz/gal × 6.236 = g/l	g/l × 0.1604 = oz/gal
lb/sq in × 0.07031 = kg/cm^2	kg/cm^2 × 14.22 = lb/sq in
lb/sq in × 6.895 = kN/m^2	kN/m^2 × 0.1450 = lb/sq in
Therms × 105.5 = MJ	MJ × 0.009478 = Therms
kWh × 3.600 = MJ	MJ × 0.2778 = kWh
Btu × 1.055 = kJ	kJ × 0.9478 = Btu
ft lb × 1.356 = J	J × 0.7376 = ft lb
ft lb × 1.356 = Nm	Nm × 0.7376 = ft lb
hp × 745.7 = W	W × 0.001341 = hp
hp × 745.7 = kW	kW × 1.341 = hp
1 unit of electricity = 1kWh = 1000 Watts (1 kW) for 1 hour	
282.5 / mpg = litres per 100 km	282.5 / (litres per 100 km) = mpg

Appendix 5

Valuation of Stock in Trade

The following calculations of length, area, and volume may be used in conjunction with the mass conversion table where it is necessary to produce estimated tonnages for valuation purposes.

Length

Circumference of a circle = 2 × pi × radius

Area

Squares and rectangles = width × length
Triangles = base × (vertical) height ÷ 2
Circle = pi × radius × radius
Cylinder (no ends) = 2 × pi × radius × length
Cylinder (with ends) = as above plus 2 × pi × radius × radius

Volume

Cubic = width × length × height
Ridge = width × length × height ÷ 2
Cylinder = pi × radius × radius × length
Pyramid = base area × vertical height ÷ 3
Cone = pi × radius × radius × vertical height ÷ 3

Notes

pi = ϖ = 22 ÷ 7, or 3.1415927
Radius = ½ diameter
10,000 m^2 = 1 ha
1,000 kg = 1 tonne
1,000 cc = 1 litre
1,000 litres = 1 m^3

Mass Conversion Table

	kg per m^3	m^3 per tonne
Wheat	776	1.3
Barley	710	1.4
Oats	510	2.0
Wheat meal	500	2.0
Barley meal	525	1.9
Oat meal	360	2.8
Maize	770	1.3
Peas	770	1.3
Beans	830	1.2
Baled wheat straw	77	13.0
Baled barley straw	88	11.4
Baled oat straw	95	10.5
Baled hay	170	6.0
Silage	720	1.4
Tower silage	800	1.25
Potatoes	650	1.5
Farmyard manure	900	1.1
Water	1000	1.0

Index